Aesthetics and the City provides an insightful and rich set of contributions into how to make sense of urban worlds. Through its eleven chapters, the edited book makes the case for a sensory approach to the studying of the production of cities. It is a lovely addition to the field of urban studies.

Professor Kevin Ward, *The University of Manchester, UK*

Aesthetics and the City teaches us that making sense of and reimagining cities requires understanding their aesthetic dimension. It is highly recommended for all urban scholars and professionals who want to explore how aesthetics intersects with urban life, influencing design visions, everyday experiences, and the broader conceptualisation of cities.

Andrea Borsari PhD, *Professor of Aesthetics, The University of Bologna, Italy*

Aesthetics and the City

Aesthetics and the City engages aesthetics to explore the role of the city in urban experience. Drawing on diverse theories and global case studies, this edited collection examines how aesthetics relates to how cities and urban spaces are perceived, organised, and transformed.

This book celebrates and ponders the wide diversity of aesthetic approaches within urban studies, noting that the way aesthetics is understood impacts what can be understood about cities and the urban order more generally. In its most general sense, aesthetics refers to our sensuous relation to the world. It invariably figures in how we make sense of the city and ourselves—bound to how urban life is experienced imaginatively, materially, socially, culturally, and politically. In an era where scholars have expressed concern at epistemological city-centrism, aesthetics is proposed as a versatile concept through which the centrality of the city to urban thought can be assessed. The book also explores how aesthetics intersects with a range of tangential concepts including power, the political, art, and affect. Ultimately, it makes the case that this diverse ensemble of approaches to aesthetics can enable scholars to understand the city and its enduring relevance to urban thought.

This book focuses on the concepts of "aesthetics" and "the city" and will appeal to scholars and students in urban studies, human geography, planning, politics, and sociology.

Joe Blakey is a Political Geographer at the University of Manchester, UK, interested in conceptualising and understanding political change, aesthetics, and environmental knowledge politics. His research is driven by the need to understand the depoliticisation of marginalised voices and perspectives, and how knowledge about the environment is produced, contested, and mobilised in response to the global climate and ecological crises.

Amy Barron is a Social and Cultural Geographer at the University of Manchester, UK, interested in the lived experience of urban difference, inequalities, and social categories. These themes are often explored through the lenses of age, ageing, and the life-course.

Routledge Studies in Urbanism and the City

Mapping Legalities
Urbanisation, Law and Informal Work
Edited by Thomas Coggin and Roopa Madhav

Urban Marginality, Racialisation, Interdependence
Learning from Eastern Europe
Edited by Filip Alexandrescu, Ryan Powell and Ana Vilenica

City Re-construction
Urban Policy Innovation Towards Sustainable Cities in MENA Region
Edited by Ali A. Alraouf, Odeh Al-Jayyousi, and Kheir Al-Kodmany

Hybrid Urbanisms in Secondary Cities of the Global South
Insights in urban planning and infrastructure delivery in Ghana and Peru
Edited by Christian Rosen and Nina Gribat

Cities: Inclusive, Liveable, and Sustainable
Edited by Ashish Kumar Srivastava and Iva Ashish Srivastava

Queerburbia: LGBTQ2S suburban place-making
Alison Bain and Julie Podmore

Building the City
Everyday Lives of Migrant Workers
Mark Jayne, Wu Siying and Wu Chenhui

Aesthetics and the City
Edited by Joe Blakey and Amy Barron

For more information about this series, please visit https://www.routledge.com/Routledge-Studies-in-Urbanism-and-the-City/book-series/RSUC

Aesthetics and the City

Edited by Joe Blakey and Amy Barron

Routledge
Taylor & Francis Group
LONDON AND NEW YORK

First published 2026
by Routledge
4 Park Square, Milton Park, Abingdon, Oxon, OX14 4RN

and by Routledge
605 Third Avenue, New York, NY 10158

Routledge is an imprint of the Taylor & Francis Group, an informa business

British Library Cataloguing-in-Publication Data
A catalogue record for this book is available from the British Library

ISBN: 978-1-032-37235-8 (hbk)
ISBN: 978-1-032-37237-2 (pbk)
ISBN: 978-1-003-33599-3 (ebk)

DOI: 10.4324/9781003335993

Typeset in Times New Roman
by SPi Technologies India Pvt Ltd (Straive)

Contents

Figures

Contributors

Taher Abdel-Ghani is an architect, urban researcher, lecturer assistant, and independent filmmaker. His academic publications explore an interdisciplinary approach to cinematic urbanism, covering other fields such as sociology and education. Some of his short films were screened in Egypt, Germany, France, Portugal, Italy, and the United States.

Amy Barron is a Social and Cultural Geographer at the University of Manchester interested in the lived experience of urban differences, inequalities, and social categories. These themes are often explored through the lenses of age, ageing, and the life-course.

Joe Blakey is a Political Geographer at the University of Manchester interested in conceptualising and understanding political change, aesthetics, and environmental knowledge politics. His research is driven by the need to understand the depoliticisation of marginalised voices and perspectives, and how knowledge about the environment is produced, contested, and mobilised in response to the global climate and ecological crises.

Julian Brigstocke is a Cultural Geographer and Social Theorist at Cardiff University. His research focuses on speculative geographies of authority and power.

Mónica Degen is a Professor of Urban Studies at Brunel University, London. She has worked on international projects researching the role sensory experiences play in framing architectural practices, urban planning, and culture in cities from Doha to Cologne, Barcelona, and London. Her latest book is *The New Urban Aesthetic: Digital Experiences of Urban Change* (Bloomsbury, 2022).

Stefanie Fridrik BA BA MA, researches transformative political education that intersects with intervening artistic practice. Since 2022, she has been a doctoral candidate at the Institute for Art Education at the University of Fine Arts (HFBK) Hamburg. In 2024, she received the DOC-team scholarship from the Austrian Academy of Sciences. Currently, Stefanie works as a research assistant, project manager, and trainer at Demokratiezentrum Wien (Democracy Center Vienna).

Günter Gassner is an interdisciplinary scholar at Cardiff University who works at the intersection of design studies, critical theory, and political philosophy. He explores relationships between politics and aesthetics, urban violence and liberatory antifascism, and visual methodologies and historiographies.

Beatriz Guijarro Turégano received her PhD from Barcelona University for her research on "Rethinking crossroads: The implementation of the Superilles model in Barcelona". She is a member of the research group LiquenDataLab, the Grup de Recerca sobre Exclusió i Control Socials (GRECS), and currently working as a project manager at Barcelona City Council.

Amr Ibrahim is a design, architecture, and art professional who specialises in branding and innovative design. With a master's from Politecnico di Milano, his award-winning work spans branding, packaging, and digital art. He has spoken at prominent design events and his work has been exhibited, featured, and published on an international scale.

Friederike Landau-Donnelly is a political theorist, urban sociologist, and cultural geographer. Friederike is interested in intersections between politics and space, and researches modalities of conflict in cultural policy and governance as well as in concrete places such as museums, monuments, and public space(s). In addition, Friederike publishes poetry as #PoeticAcademic: https://friederikelandau.com/poeticacademic/

Sanna Lehtinen is a Senior University Lecturer in History and Theory of Contemporary Art at the Transdisciplinary Art Studies unit (TAITE, prev. TAS/YoYo) at the Aalto University School of Arts, Design and Architecture (AaltoARTS). She is also a Docent in Aesthetics at the University of Helsinki. Her research focuses on the intersection of urban environmental aesthetics, philosophy of the city, and philosophy of technology.

Yara Mohamed Yara M. Ewida, an Egyptian architect and independent researcher, specialises in bio-digital architecture, focusing on blending nature and human spaces for sustainability. Awarded the Swarovski Foundation's Creatives for our Future prize in 2021, she was also a 2023 finalist in YAC Academy's "Eterna Bellezza" competition for her bio-design innovations.

Salila Vanka is an Associate Professor at the Wadiyar Centre for Architecture, Mysuru, India. Besides degrees in architecture and planning studies from India and the USA, she holds a doctoral degree in Urban and Regional Planning from the University of Michigan, Ann Arbor. Her research interests and work focus on public space, urban spatial politics and participatory planning in the Global South.

Josh Widera is a PhD candidate at the University of Southern California. His dissertation research investigates platform urbanisms and the labour and cultures underlying algorithmic urban infrastructures. His work is published in *Performance Philosophy*, the *International Journal of Communication*, and *New Media & Society*, among others.

Preface

This volume explores how aesthetics shapes our understanding and experience of the city. Drawing upon thought and empirical cases from diverse global contexts, the book proposes aesthetics as a reflexive concept through which we can consider how cities are understood, contested, and transformed, whilst also heeding warnings around the perils of city-centric thinking.

The book originated in two paper sessions we organised for the 2021 Research Committee 21 (RC21) Conference in Antwerp, which ultimately took place online amidst the unfolding global pandemic. These sessions emerged from a growing recognition of a turn towards aesthetics in urban studies and geography, alongside our desire to identify a shared conceptual ground. This was no small task, given that one of us is guided by non-representational theory's focus on liveliness and life-as-it-happens, and the other by post-foundational political theory's attention to contingency and radical difference. Yet aesthetics provided common ground—a testament to its conceptual richness and untapped potential.

This sense was reinforced as session abstracts arrived from scholars across Philosophy, Sociology, Architecture, and Geography. The decision to use aesthetics as a critical lens for thinking about the city was shaped by multiple influences: the *Sensing the City* theme of the RC21 conference, debates on city-centrism in Urban Studies, and similar discussions in Human Geography around scale. At the same time, our respective backgrounds—one rooted more in aesthetics, the other in urban thought—steered us towards the topic of aesthetics and the city.

Finally, this edited volume is a testament to the collaborative efforts of many scholars writing together under often challenging circumstances. We are deeply grateful to all contributors for their patience, generosity, and spirit of collaboration throughout this process.

Joe Blakey and Amy Barron,
Manchester, December 2024

Acknowledgements

The editors would like to thank the organisers of the 2021 Research Committee 21 (RC21) Conference, where we hosted two sessions that formed the basis of this book. Our gratitude also extends to the attendees and participants of these sessions. We also acknowledge funding from the School of Environment, Education, and Development (SEED) Early Career Strategic Research & Scholarship Fund, which supported a workshop with contributors in Manchester that contributed to the cohesiveness of this publication. Special thanks go to the workshop attendees and particularly Harry Quealy for his assistance in hosting. We are also grateful to the Cities Politics and Economies Research Group (CPERG) for their feedback on early drafts that informed the Introduction and Conclusion chapters, as well as to Kevin Ward and Sarah Marie Hall for their insightful comments on later versions of these drafts. Any remaining interpretations and errors are our own. Finally, we thank the contributors for their valuable input and patience, and the team at Routledge/Taylor & Francis for their support in bringing this book to fruition.

1 Introduction

Making Sense of the City

Joe Blakey and Amy Barron

Introduction

How do we make sense of the city? This question is deceptively simple. If you ask an everyday person on the street what a city *is*, they might circle an area on a map, point to a formalised administrative boundary, or suggest that a city is an urban centre of greater extent, population, or significance than a town or village. Given that its etymological roots are in the Latin *civitas* (as a social body of the *cives*, or citizens), one might even expect a remark that a city is simply "where many people live". There are myriad possible responses, and the longer one dwells on this question, the more David Harvey's claim that the "city is manifestly a complicated thing" rings true (1973: 22).

Cities mean different things to different people, and different cities take different shapes and forms and play different roles in individuals' lives. At the same time, for urban scholars, the city is increasingly seen to occupy an awkward conceptual space. The conceptual notion of the city has arguably fallen out of favour or, at the very least, is being decentred from much urban scholarship (Davidson and Iveson, 2015). The city, once the cornerstone of urban thought, is now often seen as a presupposition—a predetermination—an assumption about the world rather than something that *is*. It is seen as distracting or contorting our understandings of the social, political, and economic processes that happen "on the ground" and often exceed the conventional boundaries of "the city".

How, then, might the city be approached in a manner sympathetic to the many ways in which it is understood and urban studies' more recent hesitation around the concept, which is seen as at risk of dampening our appreciation for urban life as it unfolds? This question is crucial given the current impasse in urban studies regarding the city's epistemological status. Despite critiques of the concept, the city remains prominent in the popular imagination, shaping—and being shaped by—everyday lived experiences. We talk about cities, envision them differently, and make demands related to them. These dynamics factor into urban experience, despite important epistemological questions about the nature of "the city" persisting. In this edited volume, and by way of response, we propose aesthetics as a fruitful line of inquiry, as a means to

DOI: 10.4324/9781003335993-1

balance presuming too much about the city in advance and acknowledging the city's role in shaping collective imaginations and individual lives.

Indeed, the concepts of aesthetics and the city find themselves on somewhat opposing trajectories. Whilst the concept of the city is being decentred from analyses, there has been somewhat of an uptick of interest in the concept of aesthetics, particularly in urban studies and geography. For instance, scholars have considered how aesthetics lends legitimacy to planning visions (Ghertner, 2015), but also how it becomes a cornerstone for disruption and disagreement around what form the city ought to take (Gassner, 2023). At the same time, scholars have considered the everyday experience of aesthetics, be this in focusing on more phenomenological accounts, or those accounts that consider how aesthetics is encountered, thereby offering a relational understanding between subjects and objects (Hawkins and Straughan, 2016a; Wilson, 2016; Cooper, Cook, and Bilby, 2018). The field of geohumanities, bridging geography with arts and humanities scholarship, further highlights this trend. Yet, as Tursić (2019: 3) notes, aesthetics in urban studies remains "fragmented" and is often tied to sensory pleasure. In our discipline of geography, Hawkins and Straughan (2016a) sought to reaffirm the value of aesthetics, which was considered omnipresent yet under-explored.

We are certainly not alone in pondering the value of aesthetics to thinking, and potentially rethinking, the city. Tursić (2019) notes that only a handful of urban scholars have considered aesthetics in relation to urban conditions and the city. For Tursić (2019), the value of aesthetics is that it compels researchers to consider what and when the city is, a sentiment this book embraces. Also notable is the work of Virmani (2021): 3), whose edited volume on the aesthetics of urban environments has drawn attention to the "everyday aesthetic experience of the city" as an essential element of the public sphere. Amidst this aesthetic scholarship, there has been a trickle of literature considering—often implicitly—the role of aesthetics in how we experience, order, and remake the city (Cooper et al., 2018; Degen and Rose, 2024; Gassner, 2023; Ghertner, 2010; Degen, 2008).

Aesthetics is therefore a lens through which cities can be apprehended and even possibly remade; whether it be discussing what distinguishes a city's particular sensory experience, what constitutes a beautiful city, the role of "common-sense" in (re)ordering cities, the artistic and aesthetic practices in cities, or ultimately in deciphering what a city "is". As such, in an era where scholars have expressed concern at epistemological city-centrism, we ask what it means to think through an aesthetics of the city. Through a range of detailed empirical cases spanning the globe, this volume will consider how approaches to aesthetics impact understandings of the city and the issues that define them; demonstrate the role aesthetics plays in how we experience, order, and remake the city; and utilise aesthetics to make the case for the continued importance of the city to urban thought.

In the remainder of this chapter, we make the case for considering aesthetics to "make sense of the city". First, we retrace the debates across urban studies that have drawn the city into conceptual question. Second, we provide a brief genealogy of the concept of aesthetics and its use in urban thought. Finally, we outline the chapters that follow.

Questioning the City

Across the interdisciplinary field of urban studies, the question of the very concept of the city, and its role in our theoretical and political writings, is being repeatedly drawn into question. There has been an assortment of competing claims around the nature of cities and how exactly they should be studied and conceptualised (Scott and Storper, 2015). As Davidson and Iveson (2015) state, this debate—which has been simmering since the 1980s—has "reached boiling point" (Davidson and Iveson, 2015: 647). Moreover, the backdrop to all of this is that there has been a deep change to urban experience in the last few decades, where around 55% of the world's population now reside in these things we call "cities" (UN Habitat, 2022). But this is also a change that is rewriting the qualitative nature of urban life which is "unsettling many of the foundational assumptions of urban studies" (Angelo, 2017: 159). At the same time, urban thought has been charged with taking into account the vastly diverse empirical phenomena that work to compose cities (Scott and Storper, 2015).

In a 2008 editorial in Urban Geography, Kevin Ward writes that "there is a systemic weakness in much past work in the manner it has conceived of cities [...] as discrete, self-enclosed, and analytically separate objects" (2008: 407). Indeed, across numerous fields, there has been an effort to incorporate those things traditionally thought of as "outside" a city into urban research (Angelo, 2017). Take, for instance, urban political ecology, in its emphasis on "metabolic" flows of capital, people, and things beyond the traditional bounds of the city (Heynen, Kaika, and Swyngedouw, 2006; Kaika, 2004), or post-colonial thinkers, who critique how urban theory is projected and universalised from a Western developmentalist vantage point, creating a binary between the city and the "not-yet" city (c.f Robinson, 2011; Sheppard, Leitner, and Malignant, 2013; Hentschel, 2015). The point is that the pervasive "city lens" in urban thought, where cities are seen against a non-urban outside, is increasingly understood as unsuitable for the study of urban environments (Angelo, 2017: 160).

It is common to trace a genealogy of the city in urban thought to the work of the Chicago School of Urban Sociology in the early to mid-1990s (Scott and Storper, 2015). However, it is important to note that this enduring genealogy is an origin story of largely Anglophone cities, against which non-Anglophone cities were often assessed and analysed. The Chicago School understood the city as evolving in predictable patterns of expansion and succession, where specific social environments and spatial arrangements fostered particular social relationships and norms. Castells (1968), however, critiqued this, arguing that the Chicago School's questions about the city lacked an urban focus. There was little, in Castells' opinion, that was being written that could not be said of society more generally. In the 1970s, Marxist thought became the dominant way of considering the city, with the likes of Castells (1972) arguing that the Chicago School obstructed capitalism's role in organising society. For many Marxist scholars, the city was an arena for class struggle around resources and rights (Lefebvre, 1970; Harvey, 1973).

The 1980s also ushered in two further trends in thinking about and around the city, as observed by Scott and Storper (2015). First, feminist thought was influential in ushering in a consideration for the gendered aspects of cities (c.f. Massey, 1991; England, 1994; Pain, 2001; Bondi and Rose, 2003) whilst others sought to rejuvenate a focus on the role of class, race, and ethnicity in making and remaking the city (Amin, 2002; Watt, 2006). Following the millennium, they would be joined by post-colonial thinkers who, for instance, brought a shift towards thinking not just *about*, but *from*, cities in the Global South rather than the Global North (Parnell and Robinson, 2012; Roy, 2009; Simone, 2020). Second, was a turn towards considering the city in relation to the wider processes of the global urban system (Friedmann and Wolff, 1982; Sassen, 1991), an antecedent to a later turn to thinking about cities in terms of flows, processes, and networks through the likes of Deleuze and Guattari (1987[1980]) and DeLanda (2006).

In their various ways, these shifts contributed to a fragmentation of how the city was conceptualised in urban studies. One such fragment might be termed new particularism, where a strong emphasis was placed on the individual distinguishing traits of cities (Amin and Graham, 1997). This arose out of a feeling that it was futile to say anything on an abstract level about cities, because they are "so big, so complicated and so lacking in easily identifiable boundaries that any attempt to define their essential characteristics is doomed to failure" (Scott and Storper, 2015:1). Many scholars have drawn upon Actor Network Theory (ANT) and assemblage thinking to this end, focusing on the relations and practices that compose cities and the phenomena they are entwined with, giving rise to singular and situated understandings of phenomena occurring in these things we call cities (McFarlane, 2011; Brenner, Madden, and Wachsmuth, 2011).

This move towards relational, networked understandings, and the problematisation of what lies "in" and "outside" of cities and the processes they constitute and are constituted by, has ultimately led to the city falling out of favour in some quarters. Consider, for instance, the critique of methodological cityism: scholarship that assumes the "city" as a self-evident unit of analysis while automatically categorising all non-city areas as an external "rural" or not-city space (Angelo and Wachsmuth, 2015). Perhaps the greatest expression of this is the work of Brenner and Schmid (2015) who argued that we are now in an era of planetary urbanisation, where notions of both the city and the urban, which exist seemingly in relation to some non-urban hinterland (Cronon, 1991), are problematic. Here, the city and its counterpart of the urban, stand in the way of fully comprehending urban processes, and so the scholars called for a "new epistemology of the urban". The planetary urbanisation thesis is widely critiqued and debated, but is arguably symptomatic of a more widespread acceptance that much discourse is naïve in how it discusses and reifies cities (Walker, 2015).

Of course, it is important not to overdraw this as a conclusion. Scholars less concerned with these conceptual battle lines do utilise the city in more of an "anything goes" way, akin to how it is used in everyday conversation (c.f. Florida, Rodríguez-Pose, and Storper, 2023; Macrorie, Marvin, and While, 2021). This observation underlines our main point: regardless of the conceptual arguments and important epistemological cautions against assuming qualities of urban phenomena based on preconceived notions of the city, the city nonetheless occupies a significant place in academic and popular individual and collective imaginations. It remains an integral part of our discourse, our popular imaginaries, and of our political systems and visions. We propose aesthetics, then, as a way to walk the tightrope between the risks of presuming something about real-world phenomena and also taking seriously the fact that scholars and everyday folk talk about the city in a way which makes it meaningful. Aesthetics is suggested as a way to take seriously the warnings around the epistemological effects of city-centric thinking, whilst acknowledging how we nonetheless individually and collectively make sense of the city.

Aesthetics

The term "aesthetics" was coined in 1735 by Alexander Gottlieb Baumgarten, in the final pages of his master's thesis, *Meditationes philosophicae de nonnullis ad poema pertinentibus* (Philosophical meditations on poetry), written in Halle, Germany (Reiss, 1994). Unlike its popular association with beauty, Baumgarten's concept of aesthetics pertained to the nature of poetry and our experience of it. He deliberately introduced the term to bring *epistêmê* (knowledge or understanding) and *aisthetikê* (which has roots in the Greek term *aisthēsis*, what is sensed or perceived) into relation. Baumgarten's aesthetics can therefore be understood as the science of what is sensed, perceived, and known (Aschenbrenner and Holther, 1954). Over the following 15 years, he further developed these ideas, culminating in the first two volumes of *Aesthetica* (1750–1758), which are widely regarded as having instigated the study of aesthetics across philosophy. At the outset of the book, he gives the following definition:

> Aesthetics (as theory of liberal arts, as lower theory of knowledge, as art of beautiful thinking, as art of thinking analogous to reason) is the science of sensitive knowing.
>
> (Baumgarten in Euron, 2019: 59)

In *Aesthetica*, Baumgarten made the case for recognising the value of perception to cognition. He positioned aesthetics as a younger sibling to logic, with the latter seen as unable to cover all forms of cognition. This was a radical step, as it coined a whole new discipline by bridging the mind-body distinction that had become commonplace in philosophy. The intellectual pursuit of logical reasoning was often viewed in contrast to the sort of rabble received by the senses, traditionally seen as a more chaotic and disorderly faculty. Baumgarten, then, was attempting to render *aisthēsis* scientific, with a set of associated rules and truths comparable to logic. In short, it was an attempt to tackle how philosophy fetishised reason over perceptual experience to create a cohesive theory of knowledge that he termed "gnoseology" (Shusterman, 1999). This was not an argument for equality between the senses and reason—for Baumgarten referred to perception as a "lower faculty"—but rather to integrate it into the study of logic.

Baumgarten's premature death in 1762 meant he never completed the final third of *Aesthetica*. Nonetheless, the term gained significant traction, with Jean Paul Richter observing in 1804 that "[t]here is nothing more abundant in our time than aestheticians" (Richter in Herder, 2006: 2). Baumgarten's legacy, though, was overshadowed by Kant's first (*The Critique of Pure Reason*, 1781) and third (*The Critique of Judgement*, Kant, 1790) critiques, both of which addressed aesthetics. While Baumgarten is widely credited with founding the field, it is Kant who is seen as establishing the foundation for subsequent work on aesthetics. Kant's work is also highly indebted to Baumgarten's. Aside from using Baumgarten's *Metaphysica* in his moral philosophy lectures (Fugate and Hymers, 2018), Kant borrowed much of Baumgarten's terminology, including aesthetics (Kannisto, 2016).

Perhaps the greatest legacy of Kant's thought is the "wrenching duality" at the heart of aesthetics, as Deleuze later described it (Deleuze, 1990: 297). This duality, which may explain why aesthetics is such a broad and varied field today, can be understood through comparing Kant's *First* and *Third Critiques*. In his *First Critique, The Critique of Pure Reason* (Kant, 1998[1781]), Kant develops a theory of sensibility. Kant remarks that all knowledge ultimately stems from experience (Kant, 1998[1781]). For Kant, "experience" is complex, for it relates to an interaction between mental concepts and sensibility, which he defines as "the capacity of producing representations through the way in which our senses are affected by objects" (Nuzzo, 2006: 582). A key focus of the *First Critique* is on *a priori* conditions that enable experience (Dikeç, 2012). Kant caveats that there is a "transcendental aesthetic", where space and time are only "pure" concepts built into our minds that provide "form to our experience of the world" yet do not arise themselves from experience (Dikeç, 2012, p. 269). Alongside other concepts from the "empirical aesthetic" of encountered sensations, we derive a spatio-temporal system of relations that give conceptual form to the perceptions of our senses (Dixon, Hawkins, and Straughan, 2012; Dikeç, 2015).

In his *Third Critique, The Critique of Judgement*, Kant shifts his focus to aesthetic *judgements*, considering declarations of beauty, taste, and awe-inspiring experiences of "the sublime" (Shaviro, 2007). Central to this discussion is the idea of *sensus communis*, or "common-sense". For Kant, aesthetic judgements always involve a presupposition that they can be universally communicated and that universal agreement is possible (Dikeç, 2012). For example, when one declares something to be "beautiful", there is an implicit assumption that everyone should share the same feeling:

> Whenever we make a judgement declaring something to be beautiful, we permit no one to hold a different opinion [...] we regard this underlying feeling as a common rather than as a private feeling. But if we are to use this common sense in such a way, [...] it does not say that everyone will agree with my judgement, but that he [sic] ought to. [Common sense...] is a mere ideal standard. With this standard presupposed, we could rightly turn a judgement that agreed with it, as well as the liking that is expressed in it for some object, into a rule for everyone.
>
> (Kant, 1790: 89 emphasis added)

Sensus communis, then, is the expectation that aesthetic judgements, like declaring something beautiful, appeal to a *shared* sensibility (Kant, 1790). Between his *First* and *Third Critiques*, Kant explores aesthetics from two perspectives: an empiricist perspective, where aesthetics is closely tied to subjective feelings and affections beyond cognition (*aisthēsis*), and a more rationalist perspective, where aesthetics involves cognitive mediation of an object in a manner that is universalisable (*sensus communis*) (Dixon et al., 2012). This dual approach has led to a division in how aesthetics is understood: between the socially constructed norms by which we judge things (what has artistic merit, what is beautiful, sublime, and so forth) and "the sensuous exploration of subjects, bodies and spaces through a focus on experiences that are in excess of rational thought" (Hawkins and Straughan, 2016a: 8; Deleuze, 1990). This duality is also why Eagleton (1990: 28) would describe aesthetics as a "contradictory, double-edged concept".

Aesthetics in Urban Thought

In urban studies, as in geography, aesthetics has been present but often overlooked (Hawkins and Straughan, 2016a). While geography has increasingly embraced aesthetics, the concept has had less traction in urban studies. For this reason, Tursić has called for a more serious engagement with aesthetics from urban studies, observing that "aesthetics seems present in urban studies, but only in fragments [...] and [it] is usually equated with sensuous pleasures, which amputates the critical element from aesthetic theory" (Tursić, 2019: 206). Whilst we disagree that more sensory forms of exploration (*aisthēsis*) necessarily remove critique, Tursić rightly highlights aesthetics as an underexplored aspect of urban thought.

Perhaps one explanation for the historical oversight of aesthetics in urban studies is that, as Engelmann and McCormack suggest, aesthetics is often viewed as "some kind of superficial veneer layered over more important or substantive issues", neglecting how, "as a domain of sense making, the aesthetic is critical to how material dynamics of the world are sensed and take shape as matters of concern" (Engelmann and McCormack, 2018: 242). Aesthetics in urban studies, then, has been too often "regarded as a rather abstract, esoteric enquiry" (Taylor, 2003: 1610). One might speculate that this is fuelled by the popular, siloed imaginaries of aesthetics as relating to beauty, art, representation, and appearance, rather than on the broader social, cultural, and political implications of the term.

However, this lack of engagement with aesthetics should not be overstated, and this chapter now turns to explore how it has nonetheless been addressed in urban thought. What is notable across these studies is the variety of ways aesthetics is understood and deployed, revealing different aspects of cities and urban processes more generally (Murakami Wood and Abe, 2011: 3242). For instance, the millennium saw a "renewed focus on people's sensory experience" of cities, such as Taylor's (Taylor, 2003: 1610) work on the aesthetic experience of car traffic and how it has become synonymous with the experience of cities more generally. A more recent example is Edensor's (2015) study on the aesthetics and atmospherics of darkness and shadow, emphasising how light and dark shape the nocturnal city experience, affecting spaces, atmospheres, movement, and activities, from shopping to stargazing.

Alternatively, one might consider how the early 2000s saw a focus on the aesthetics of urban regeneration. Ley (2003) examined the role of artists in gentrification, while Leszczynski and Kong (2023: 773) described gentrification as "a process of aesthetic change", critiquing how images of decay fuel gentrification. Julier (2005), meanwhile, considered how design-led urban regeneration goes beyond buildings and brochures, wherein a network of different signifiers contribute towards creating a "designscape". A recurring theme amongst such literature is the branding and lived experience of gentrification projects, such as in the work of Degen and Rose (2024), who explore how aesthetics is mediated by digital branding and the differential ways this affects bodies.

In recent years, there has been a growing exploration of politics and political change through the lens of aesthetics. In other words, an aesthetic turn in political thinking has emerged, which has spread to urban studies (Kompridis, 2014). As Yusoff (2010: 77) argues, "aesthetics must be considered as part of the practice of politics; a space where things are made, both materially and semiotically (to paraphrase Haraway) and a space that configures the realm of what is possible in that politics".

Scholars have frequently built upon the work of Jacques Rancière, who insists "politics is aesthetic in principle" (Rancière, 1999: 58). Rancière, offering a twist on both components of Kant's aesthetics, suggests that we are governed by aesthetics as a form of common-sense—a normalised way of making sense of things. This, he refers to as "the police order" or "the distribution of the sensible", which determines the places, roles, and entitlements within society (Dikeç, 2012). Rancière, however, maintains that this ordering is always imperfect, leading to inequalities and exclusions. These "parts of no part" can always "declare a wrong", prompting a "redistribution of the sensible" in a political moment that reconfigures common sense (Rancière, 1999).

Others have turned to Hannah Arendt, who sought to develop a theory of politics from Kant's *Critique of Judgement* (Sjöholm, 2015), resulting in a political approach to aesthetics centred on the creation of a common world (Dikeç, 2012). For Arendt, politics is rooted in the idea of freedom and plurality. Freedom, for Arendt, is an inherent human ability to "make a beginning", to initiate action, and to intervene in the status quo through politics (Gassner, 2023). Plurality is also central because human beings are fundamentally different, yet each possesses an equal capacity to engage with and disrupt the public realm in unique ways (Gassner, 2023; Dikeç, 2012). Aesthetics underpins these interventions because, first, politics is fundamentally about "appearances and the reception of appearances in the public realm" (Gassner, 2023: 36) and, second, because political action is perceived, heard, and understood through the senses, where judgement and imagination arise as functions of sensibility (Sjöholm, 2015).

In addition to these so-called "post-foundational" articulations of aesthetics (Marchart, 2007), Walter Benjamin's work has animated discussions of aesthetics and politics within urban studies and cognate fields (Gassner, 2019). Benjamin's diverse writings span art, mechanical reproduction, memory, modernity, technology, fascism, and the city, reflecting his critique of cultural and social issues of his era. In 1930s Germany, Benjamin developed his arguments against the aestheticisation of politics, where fascism used spectacle to manipulate and control the masses, diverting focus from rectifying social inequalities. Instead, Benjamin championed the politicisation of art for revolutionary purposes, seeing it as a tool for social and political transformation. With the rise of mass culture, Benjamin cautioned that mass-produced art could be used as a tool for both liberation and domination.

Many studies drawing upon politically inflected approaches to aesthetics are articulated either around the city as the stakes of politics or the city as the site through which such aesthetic politics can (or cannot) unfold. Gassner (2023), for instance, argues that democratising the cityscape requires an aesthetic approach that goes beyond visuals to include shared, common-sense perceptions of the city's form. Elsewhere, Swyngedouw (2009), critiques the "post-political" city as resistant to political and aesthetic change. Tedesco and Davies (2022) advocate viewing the politics of the city as an aesthetic subject to challenge the perception of cities as part of a nested hierarchy from global to local.

Asher Ghertner's work on aesthetic governmentality also reflects how aesthetics, and its role in power configurations, is finding a foothold in urban studies. Ghertner defines aesthetic governmentality as an "imposed aesthetic order" that guides how a population should behave (2010: 208). Ghertner builds on Foucault's concept of governmentality, developed in the late 1970s in his lectures at the *Collège de France* which focused on how individuals' and populations' conduct might be shaped by things like statistics. Ghertner (2015) examines how aesthetics instead became a key tool for governance in Delhi, promoting a "world-class" vision and shaping citizens to embody it. Aesthetic governmentality is also applied in Montefrio, Lee, and Lim's (2021) work on Singapore's community gardens and Pow's (2018) research on eco-cities in China, where it works to create a sense of harmony.

Some scholars have introduced prefixes to "aesthetics" to explore complementary concerns within urban studies, such as *kinaesthetics* and *anaesthetics*. Kinaesthetics combines the Greek *kinein*, meaning "to move", and aesthetics, and is defined by Latham and Layton (2020: 857) in their work on sport and fitness in city environments as "an awareness of the body in motion and proprioception", the latter is an awareness of the body's location, movements, and actions. Bissell (2022), meanwhile, develops "anaesthetics" to study gig workers in the digital economy, where concealment, projection, and resignation serve as anaesthetic tactics to protect them from being affected by their otherwise depleting experiences. As such, anaesthetics relates to the ways in which bodies are unaffected. Both studies therefore explore the intersection of aesthetics and affect—the capacity to affect and be affected (Barron and Blakey, 2023). For example, Latham and Layton (2020) examine how affects are shaped by environments and by the ways practices evolve over time.

Dixon et al. (2012), writing about human geography, note that much of aesthetic thought has historically been human-centred. However, more recent works critique this anthropocentric perspective, instead considering aesthetics as a complex interplay of sensory experiences and judgements shaped by the differential capacities of a broad range of human and non-human influences and forces (Brigstocke and Gassner, 2021; Engelmann and McCormack, 2018; Dixon, 2009; Yusoff, 2015). For example, Dixon et al. (2012) offer a post-human aesthetic approach to artworks, highlighting the constitutive role of matter in creating new worlds.

Simultaneously, much aesthetic and political thought has been Eurocentric, often grounded in Kantian traditions. However, scholars are increasingly exploring non-European perspectives (Brigstocke and Gassner, 2021). One notable avenue is the work of Édourad Glissant, a French Caribbean post-colonial thinker, whose most celebrated work *Poetics of Relation* ("*Poétique de la Relation*") (Glissant, 1990) works through questions of aesthetics, politics, and identity in relation to post-colonial experiences in the Caribbean (Brigstocke and Gassner, 2021; Jackson, 2016; Hoover, 2021). Central to Glissant's thought is the idea of "Relation" and "opacity". The former, as Glissant (1990): 11) notes, highlights how "each and every identity is extended through a relationship with the Other", while the latter, in contrast to transparency, advocates an appreciation for complexity and so better respects difference (Jeyasingh, 2024; Simone, 2016). Glissant's work defies easy summarisation and is often misread (see Jeyasingh, 2024). Nonetheless, Brigstocke and Gassner (2021) have argued that his approach can broadly be understood as an aesthetics of disruption, focused on creating new relations by disrupting existing ways of making sense, making his approach "poetic" in the sense of *poesis*, meaning "to create"

(Brigstocke and Gassner, 2021; Nassar, 2021). Hoover (2021: 472), for instance, uses concrete as a metaphor to understand exploitation, neoliberal governance, climate change, and inequality in European cities and to highlight how "urban contexts are also connected to archipelagos, landscapes and riverbeds". Alternatively, Brigstocke (2021), also drawing upon Glissant, uses sand poetically to disrupt the aesthetics of how we think about Hong Kong. Brigstock does this playfully, experimenting with different writing forms to foreground new relations.

Tangential Concepts

While aesthetics is increasingly emphasised in urban studies, we also highlight a range of tangential concepts and studies that, while not explicitly travelling under the concept of "aesthetics", operate within or alongside its broad remit of sense-making. Aesthetics, as is evident from the present discussion, is a broad church, relating most generally to sensory perception, experience, and judgement. Urbanists often engage with these themes, even if aesthetics remains implicit.

Many projects work within the remit of aesthetics in terms of perception or sensation—as *aisthēsis*—but are instead labelled as "sensory urbanism" (Jaffe et al., 2020). Sensory urbanism is an emerging research focus that considers the "embodied sensations that city dwellers experience, and the political economic effects of these experiences" (ibid: 1017). These studies explore how sensory regimes are maintained and reproduced through the wider sensorium. Whilst sensing goes beyond the "five senses" (e.g. chronoception: the sense of the passage of time), studies frequently engage vision—such as Cooper et al.'s (2018) study of sex work in Blackpool; sound—such as Atkinson's (2007) study of the "sonic ecology" of streets; smell—such as Manalansan's (2006) study of migrant lives in global cities; and taste—such as Rhys-Taylor's (2013) discussion of class, ethnicity, and generational (dis)taste towards a London seafood stand.

Another pertinent concept is "atmosphere". In the conclusion to *Geographical Aesthetics*, Hawkins and Straughan (2016c) note the significance of atmosphere in relation to aesthetic thought but observe its conceptual underdevelopment. To address this, they reference Anderson (2009), who defines atmospheres as:

> collective affects that are simultaneously indeterminate and determinate. Affective atmospheres are a class of experience that occur before and alongside the formation of subjectivity, across human and non-human materialities, and in-between subject/object distinctions.
>
> (2009: 78)

Atmospheres are shared but differentially experienced, shaping moods, feelings, and behaviours, though they are not reducible to them. For Hawkins and Straughan (2016a), considering aesthetics as atmospheres facilitates a consideration of how a diffuse set of relations can work to sustain or disrupt sensory regimes over space, bringing about differentially lived but collectively experienced affective and sensory happenings. Gandy (2017) finds salience in atmosphere for urban thought and, amongst other things, notes its potential to challenge sensory hierarchies and ocularcentrism. Atmospheres are also gaining increasing influence in urban thought, such as Degen and Lewis' (2020) study of London's Smithfield market, where sensory and temporal interactions create unique atmospheric experiences.

Atmospheres, much like Bissell's work which considered anaesthetics as being unaffected, also point to the tangential importance of affect to the study of aesthetics. For such studies, and as equating atmospheres and aesthetics suggests, aesthetics (implicitly or explicitly) plays a role in the production of affects, and so draws attention to what aesthetics *does* (Hawkins and Straughan, 2016b). Moreover, if atmospheres are something encountered, then integral to understanding aesthetics in this manner is the concept of encounter, as meetings that "make a difference [...and] enact a shift in sense perception" (Wilson, 2017: 458).

The study of representations, their social construction, and their effects also provide valuable insights (Barron and Blakey, 2023). Take, for instance, Pinder's (2013) examination of 20th-century utopian city visions and Usher's (2023) exploration of how design principles from the decorative arts translate to urban planning. One might also look at the discursive turn in urban studies around the millennium (Lees, 2004). Though, like aesthetics, the definition of discourse can be varied and slippery, it can broadly be understood as a shared system of ideas and concepts that shapes meaning, social norms, and identities. Whether through Gramsci's view of discourse as "an instrument of hegemony" or Foucault's idea of discourse-producing regimes of truth—where "language, knowledge, and power are interconnected" (Lees, 2004: 102)—discourse operates, like aesthetics, between individual expression, representation, common-sense understandings, and power (Parker, 2000; Butler, Schafran, and Carpenter, 2018). The connection is perhaps unsurprising, given Eagleton's (Eagleton, 1988: 327) claim that "aesthetics is born as a discourse of the body".

Early urbanists like Georg Simmel (1950[1903]) implicitly addressed aesthetic concerns, particularly through his concept of the blasé attitude. In *The Metropolis and Mental Life*, Simmel explores how the urban environment shapes individuals' sensory experiences, where a bombardment of stimuli results in a blasé attitude: a sense of indifference to defend themselves from sensory overload. Notably, this is a more pejorative understanding of insulating oneself from sensory stimuli than Bissell's (2022) discussion of anaesthetics. Siegfried Kracauer, influenced by Simmel and his lectures at the Königliche Technische Hochschule zu Berlin (Gilloch, 2015), explored how cultural products, particularly film, both reflect and shape their contexts. Kracauer, like Benjamin, pointed to cinema as holding a significant role in urban life in the 1920s and 1930s, "revealing the new conditions of life [but also providing] a form of sensory training for modern city living" (McQuire, 2008: 67).

Movement has been another area of urban focus on sensory experience. Rooted in the transformations of 19th-century Paris, the concept of the *flâneur* describes a detached yet engaged observer of the city who is absorbed and delights in all of its sensory experiences and minutiae (Latham, 2009). Charles Baudelaire was a champion of the *flâneur*, and the idea was further expounded upon by Benjamin. *Flâneurie* has fallen out of popularity on the streets of Paris. However, as Latham suggests, the "*flâneur* continues to have a significant place in much contemporary writing on the contemporary city" (2009: 192). Whilst the *flâneur* often represented a privileged male figure, contemporary research seeks to include a much broader spectrum of urban experiences. Meanwhile, the *dérive* or drift, introduced by the Situationists International (SI)—an avant-garde movement of social and political theorists, artists, and intellectuals active in the mid-20th century—and Guy Debord in particular, is a more purposeful act of drifting through the city (Arnold, 2019). It emphasises the importance of spontaneous unstructured explorations and, in doing so, seeks to disrupt habitual perceptions revealing the hidden dimensions of cities. Pyyry (2019) employs this approach in her non-representational urban research on spatial justice, using *dérives* and photography to challenge fixed notions of self and city, emphasising their continuous co-emergence.

Across concepts such as affect, atmosphere, dérive, discourse, encounter, the flâneur, cinema, representation, sensory urbanism, and Simmel's blasé attitude, we see a range of existing urban thought that implicitly engages with many of the aesthetic themes discussed in this chapter. Additionally, earlier sections of this chapter have already addressed more explicit tangential concepts to aesthetics such as power, politics, art, culture, the sublime, poetics, and subjectivity. This list of tangential concepts is by no means exhaustive. Additional avenues for exploration include aspects of urban aesthetics such as decay (Pohl, 2022), ruination (DeSilvey and Edensor, 2013), urban beauty (Gassner, 2023), and cities "world-class" status (Ghertner, 2015). Notably, several of these concepts also engage with creativity and creative methods, which we examine in greater detail in the conclusion.

Aesthetics as a Response to City-Centrism

Urban thought on the role and epistemological status of the city might be argued to be in somewhat of an impasse. On the one hand, the city, much like scale before it (Blakey, 2021), is seen to weigh on our thought, where the "city lens" is seen to impose an unnecessary scaffold on how urban phenomena are apprehended. On the other hand, society most generally nonetheless talks about cities. Even if the physical boundaries of a city are not clearly defined, and the activities related to the city extend beyond these limits, cities still play a crucial role in shaping our collective and individual imaginations. This influence is evident in the material and discursive artefacts that originate from cities—from speeches or books around them to artworks—which can transform or challenge our perceptions of cities along with reinforcing them. At the same time, though, different urban encounters might work to affect or be affected by common-sense ideas around the city.

In aesthetics, then, we find a way to walk this tightrope between taking seriously the warning around presuming something about the city in advance and the city nonetheless figuring in our collective imaginations and, therefore, our lives. We see aesthetics as a versatile concept through which the centrality of the city to urban thought can be assessed and reaffirmed. Some may be tempted to label aesthetics as yet another chaotic concept, with extreme divergence in how it is understood and applied (Marston, Jones, and Woodward, 2005). At the same time, this chapter has pointed to how aesthetics neatly intersects with a range of tangential notions—be it power, the political, art, affect, atmospheres, encounters, representation, discourse, or subjectivity—and some might argue that such notions allow scholars to express much of the same, and often in a more precise manner. Why, then, do we suggest that aesthetics, with its manifold meanings, is a fruitful notion through which to (re)consider the city?

As this edited collection seeks to demonstrate, we find aesthetics to be a means through which scholars can reflexively ask how we make sense—or ought to make sense—of the city. In emphasising our sensuous relationship with the world, aesthetics provides a lens through which cities can be apprehended. This includes discussions on what distinguishes a city's particular sensory experience, what makes cities beautiful, the role of "common-sense" in (re)ordering cities, artistic and aesthetic practices within urban spaces, and ultimately, understanding what a city "*is*". Whilst, for instance, tangential notions such as discourse might themselves speak to the power of representations or seemingly common-sense narratives around cities, they do so from specific theoretical and epistemological lineages. Aesthetics, instead, becomes a common ground for those scholars coming from different lineages who might not otherwise cross paths.

The conversations that shaped this book revealed not a chaotic concept, but a mutual interest in how we make sense of this world. Amid critiques of city-centric thinking, aesthetics emerges as a versatile concept through which the significance of the city to urban thought can be evaluated and reaffirmed. In the chapters that follow, then, we will celebrate and ponder the diversity of aesthetic approaches within urban studies, noting that the way aesthetics is understood has a vital impact on what can be understood about cities and the urban order more generally (Murakami Wood and Abe, 2011). Ultimately, we make the case that it is this diverse ensemble of approaches that facilitates a better understanding of cities.

One of the exciting elements of an edited volume is its flexibility, allowing readers to engage with the content in the order they prefer. The narrative presented in this book does not build sequentially but rather consists of independently written chapters, each offering distinct insights and perspectives on aesthetics and the city. After reading the introduction, readers are encouraged to begin with the chapters that pique their interest, and then review the conclusion. Each chapter sheds light on the interplay between aesthetics and the city but approaches the subject from different disciplinary, geographical, theoretical, and epistemological vantage points. By examining diverse case studies across the globe and spanning various points in time, the book paints a multifaceted picture of how aesthetics and the city interrelate. Given that aesthetics has proven to be such a "lively" concept with innumerable faces (Hawkins and Straughan, 2016a), this collection could never be definitive. Instead, it should be considered as a provocation. It is also in this spirit of openness through which we interpret what constitutes urban thought: as having no hard boundaries but having a remit relevant to urban questions and ideas.

If this introduction has been a provocation to prompt urban scholars concerned with the city to think through philosophies of aesthetics, then Chapter 2 seeks to invert this premise. Sana Lehtinen instead focuses on the way in which philosophical urban aesthetics, which has condensed into a field of philosophical thought over the last few decades, takes the city into consideration.

In Chapter 3, and against a backdrop of intensified digital mediation, Degen and Guijarro explore the relationship between urban aesthetics and power through branding and sensory conviviality in London and Barcelona. Using Böhme's concept of "aesthetic economy", they highlight how sensory-experiential aspects of urban life shape urban imaginaries. The "Culture Mile" in London illustrates how future regeneration of public spaces is driven by a branding strategy that emphasises sensory engagement in digital public life while recent urban branding campaigns in Barcelona show how promoting a new city brand evokes specific sensory relations in everyday public life.

In Chapter 4, Widera considers aesthetics as a sensory experience through two vignettes of bicycle delivery work. Understanding aesthetics as a form of minor theory (Katz, 1996), the chapter explains how bicycle messengers generate meaning and knowledge in their environment. It demonstrates the potential of aesthetics to foreground embodied knowledge and experience, explaining how individuals develop their unique senses of place and self. Widera shows how bicycle messengers produce a sensory experience of their lifeworld through their embodied practice of delivery work, despite its precarity, alienation, and danger.

In Chapter 5, Landau-Donnelly and Fridrik draw on empirical insights from a case study of street art and graffiti governance in Linz, Austria. Influenced by Ghertner's (2010) Foucauldian-inspired account of the "aesthetic ordering of the city", Millie's (2017) Rancière-inflected understanding of aesthetic orders as inherently political, and Marchart's (2019) term conflictual aesthetics, they argue that cleanliness policies can enforce a narrow aesthetic vision. The authors advocate for an open-ended approach to aesthetics, emphasising the importance of creative, artistic, and political expressions in shaping collective and individual understandings of art and whose opinions on aesthetics matter.

In Chapter 6, Gassner discusses the relations between borders, freedom, and imagination in the racist city, focusing on Berlin and Bristol in the 1930s and 2020s. The chapter develops a distinction between an *unbounded* aesthetic that is linked to a destructive fascist creativity and an aesthetics of *un-bordering*, which relates to a radical imagination that is bounded by racialised life-worlds. Gassner also plays with the written form to make his point, employing a non-organismic style as a critical and creative intervention against violent urban imaginaries, oscillating between two places and times.

In Chapter 7, Abdel-Ghani, Mohamed, and Ibrahim examine the East Side Gallery in Berlin and the Mohamed Mahmoud Wall in Cairo, two sites where street graffiti connects the modern city to its political past. Using Deleuze's concept of the "haptic", they explore urban aesthetics as a means of evoking cultural memory, arguing that these sites serve as socio-political spaces that reinforce aesthetics as a sensory reflection of both historical and contemporary issues. The authors demonstrate how identity and history are represented in the spatial creation of the city's image.

In Chapter 8, Blakey considers the parallels and links between urban studies' warning around city-centrism and human geography's scale debate, with the latter being oriented around how pre-existing ideas around scale might impede an understanding of phenomena as they actually occur. Blakey suggests that drawing upon a Rancièrian aesthetics provides a means to move beyond the concern of both debates, by focusing on those moments when disagreements occur specifically around the city-scale.

In Chapter 9, Brigstocke explores an Arendtian approach to aesthetics, particularly Arendt's concept of a "sixth sense" or "community sense". Using experimental urbanism projects along the Tijuana–San Diego border as case studies, and an Arendtian analysis of cities, architecture, and aesthetics, Brigstocke argues that these experimental cross-border urbanisms foster a "reality" sense by embracing a culture of exposure to plurality. In essence, Brigstocke highlights how an aesthetics of the city provides more than just a sensory analysis.

In Chapter 10, Vanka investigates collaborative public wall art projects in Bangalore, focusing on their role as a form of public engagement that intersects with political debate and questions of social and spatial justice. It suggests that these art projects often serve to manage urban space through aesthetic means, rather than genuinely engaging with deeper political or social inequities.

Finally, in Chapter 11, the book concludes by highlighting how the various approaches to aesthetics presented in this book allow scholars to critically consider what the city means without falling into the alleged traps of epistemological city-centrism. It argues that there is no single approach to aesthetics. Instead, the diversity of approaches, their neighbouring concepts, and underpinning theoretical persuasions foster a productive and critically reflective discussion of the city and how we make sense of it within urban studies. It urges an agonistic approach to considering the aesthetics of the city that values ongoing disagreement around it as productive. The book closes by suggesting six interrelated avenues for future research on aesthetics and the city.

References

Amin, A. (2002). Ethnicity and the multicultural city: Living with diversity. *Environment and Planning A*, 34(6), pp. 959–980.

Amin, A. and Graham, S. (1997). The ordinary city. *Transactions of the Institute of British Geographers*, 22(4), pp. 411–429.

Anderson, B. (2009). Affective atmospheres. *Emotion, Space and Society*, 2(2), pp. 77–81.

Angelo, H. (2017). From the city lens toward urbanisation as a way of seeing: Country/city binaries on an urbanising planet. *Urban Studies*, 54(1), pp. 158–178.

Angelo, H. and Wachsmuth, D. (2015). Urbanizing urban political ecology: A critique of methodological cityism. *International Journal of Urban and Regional Research*, 39(1), pp. 16–27.

Arnold, E., (2019). Aesthetic practices of psychogeography and photography. *Geography Compass*, 13(2), p. e12419.

Aschenbrenner, K. and Holther, W.B. (1954). Introduction. In *Reflections on Poetry*, edited by Baumgarten, A.. Berkeley: University of California Press.

Atkinson, R. (2007). Ecology of sound: The sonic order of urban space. *Urban Studies*, 44(10), pp. 1905–1917.

Barron, A. and Blakey, J. (2023). Representation/al. In *Concise Encyclopedia of Human Geography*, edited by L. Lees and D. Demeritt. London: Edward Elgar Publishing Ltd. Ch. 66.

Bissell, D. (2022). The anaesthetic politics of being unaffected: Embodying insecure digital platform labour. *Antipode*, 54(1), pp. 86–105.

Blakey, J. (2021). The politics of scale through Rancière. *Progress in Human Geography*, 45(4), pp. 623–640.

Bondi, L. and Rose, D. (2003). Constructing gender, constructing the urban: A review of Anglo-American feminist urban geography. *Gender, Place and Culture: A Journal of Feminist Geography*, 10(3), pp. 229–245.

Brenner, N. and Schmid, C. (2015). Towards a new epistemology of the urban?. *City*, 19(2–3), pp. 151–182.

Brenner, N., Madden, D.J. and Wachsmuth, D. (2011). Assemblage urbanism and the challenges of critical urban theory. *City*, 15(2), pp. 225–240.

Brigstocke, J. (2021). The aesthetics of sand: Reclaiming Hong Kong's unsettled grounds. *GeoHumanities*, 7(2), pp. 370–390.

Brigstocke, J. and Gassner, G. (2021). Materiality, race, and speculative aesthetics. *GeoHumanities*, 7(2), pp. 359–369.

Butler, A., Schafran, A. and Carpenter, G. (2018). What does it mean when people call a place a shithole? Understanding a discourse of denigration in the United Kingdom and the Republic of Ireland. *Transactions of the Institute of British Geographers*, 43(3), pp. 496–510.

Castells, M. (1968). Y a-t-il une sociologie urbaine?. *Sociologie du travail*, 10(1), pp. 72–90.

Castells, M. (1972). *The Urban Question*. Cambridge, MA: MIT Press.

Cooper, E., Cook, I.R. and Bilby, C. (2018). Sex work, sensory urbanism and visual criminology: Exploring the role of the senses in shaping residential perceptions of brothels in Blackpool. *International Journal of Urban and Regional Research*, 42(3), pp. 373–389.

Cronon, W. (1991). *Nature's Metropolis: Chicago and the Great West*. New York: W.W. Norton & Company.

Davidson, M. and Iveson, K. (2015). Beyond city limits: A conceptual and political defense of 'the city' as an anchoring concept for critical urban theory. *City*, 19(5), pp. 646–664.

Degen, M. (2008). *Sensing Cities: Regenerating Public Life in Barcelona and Manchester*. London: Routledge.
Degen, M. and Lewis, C. (2020). The changing feel of place: The temporal modalities of atmospheres in Smithfield Market, London. *Cultural Geographies*, 27(4), pp. 509–526.
Degen, M. and Rose, G. (2024). Conceptualising aesthetic power in the digitally-mediated city. *Urban Studies*, 61(11), pp. 2176–2192.
DeLanda, M. (2006). *A New Philosophy of Society: Assemblage Theory and Social Complexity*. London: Continuum.
Deleuze, G. (1990[1969]). *The Logic of Sense*. New York: Columbia University Press.
Deleuze, G. and Guattari, F. (1987[1980]). *A Thousand Plateaus: Capitalism and Schizophrenia*. Translated from French by B. Massumi. Minneapolis: University of Minnesota Press.
DeSilvey, C. and Edensor, T. (2013). Reckoning with ruins. *Progress in Human Geography*, 37(4), pp. 465–485.
Dikeç, M. (2015). *Space, Politics and Aesthetics*. Edinburgh: Edinburgh University Press.
Dikeç, M. (2012). Politics is sublime. *Environment and Planning D: Society and Space*, 30(2), pp. 262–279.
Dixon, D.P. (2009). Creating the semi-living: On politics, aesthetics and the more-than-human. *Transactions of the Institute of British Geographers*, 34(4), pp. 411–425.
Dixon, D., Hawkins, H. and Straughan, E. (2012). Of human birds and living rocks: Remaking aesthetics for post-human worlds. *Dialogues in Human Geography*, 2(3), pp. 249–270.
Eagleton, T. (1988). The ideology of the aesthetic. *Poetics Today*, 9(2), pp. 327–338.
Eagleton, T. (1990). *The Ideology of the Aesthetic*. Cambridge: Blackwell.
Euron, P. (2019). *Aesthetics, Theory and Interpretation of the Literary Work*. Netherlands: Brill.
Edensor, T. (2015). The gloomy city: Rethinking the relationship between light and dark. *Urban Studies*, 52(3), pp. 422–438.
England, K.V. (1994). Getting personal: Reflexivity, positionality, and feminist research. *The Professional Geographer*, 46(1), pp. 80–89.
Engelmann, S. and McCormack, D. (2018). Elemental aesthetics: On artistic experiments with solar energy. *Annals of the American Association of Geographers*, 108(1), pp. 241–259.
Florida, R., Rodríguez-Pose, A. and Storper, M. (2023). Critical commentary: Cities in a post-COVID world. *Urban Studies*, 60(8), pp. 1509–1531.
Friedmann, J. and Wolff, G. (1982). World city formation: An agenda for research and action. *International Journal of Urban and Regional Research*, 6(3), pp. 309–344.
Fugate, C.D. and Hymers, J. (eds.) (2018). *Baumgarten and Kant on Metaphysics*. Oxford: Oxford University Press.
Gandy, M. (2017). Urban atmospheres. *Cultural Geographies*, 24(3), pp. 353–374.
Gassner, G. (2019). *Ruined Skylines: Aesthetics, Politics and London's Towering Cityscape*. London: Routledge.
Gassner, G. (2023). Aestheticizing the beautiful city: Democratic politics and design review. *Urban Geography*, 44(6), pp. 1098–1116.
Ghertner, D.A. (2010). Calculating without numbers: Aesthetic governmentality in Delhi's slums. *Economy and Society*, 39(2), pp. 185–217.
Ghertner, D.A. (2015). *Rule by Aesthetics: World-class City Making in Delhi*. Oxford: Oxford University Press.
Gilloch, G. (2015). *Siegfried Kracauer*. Cambridge: Polity Press.
Glissant, É. (1990). *Poétique de la Relation*. Paris: Gallimard.
Harvey, D. (1973). *Social Justice and the City*. Georgia: University of Georgia Press.

Hawkins, H. and Straughan, E. (2016a). Introduction: For Geographical Aesthetics. In *Geographical Aesthetics*, edited by H. Hawkins and E. Straughan. London: Routledge. pp. 1–17.
Hawkins, H. and Straughan, E. (2016b). Section Two—Aesthetic Encounters: For Geographical Aesthetics. In *Geographical Aesthetics*, edited by H. Hawkins and E. Straughan. London: Routledge. 105–120.
Hawkins, H., and Straughan, E. (2016c). Conclusion: Reimagining Geoaesthetics. In *Geographical Aesthetics*, edited by H. Hawkins and E. Straughan. London: Routledge. 283–298.
Herder, J. G. (2006). *Selected Writings on Aesthetics*. Princeton: Princeton University Press.
Hentschel, C. (2015). Postcolonializing Berlin and the fabrication of the urban. *International Journal of Urban and Regional Research*, 39(1): 79–91.
Heynen, N., Kaika, M. and Swyngedouw, E. (eds.) (2006). *In the Nature of Cities: Urban Political Ecology and the Politics of Urban Metabolism*. Abingdon: Taylor & Francis.
Hoover, E.M. (2021). Poetic commoning in European cities–Or on the alchemy of concrete. *GeoHumanities*, 7(2), pp. 464–474.
Jackson, M. (2016). Aesthetics, politics, and attunement: On some questions brought by alterity and ontology. *GeoHumanities*, 2(1), pp. 8–23.
Jaffe, R., Dürr, E., Jones, G.A., Angelini, A., Osbourne, A. and Vodopivec, B. (2020). What does poverty feel like? Urban inequality and the politics of sensation. *Urban Studies*, 57(5), pp. 1015–1031.
Jeyasingh, T.E., (2024). Édouard Glissant and the importance of reading well: Opacitic-reading as geographic method. *Transactions of the Institute of British Geographers*. https://doi.org/10.1111/tran.12685 (Accessed 1st January 2025).
Julier, G., (2005). Urban designscapes and the production of aesthetic consent. *Urban Studies*, 42(5–6), pp. 869–887.
Kaika, M. (2004). *City of Flows: Modernity, Nature, and the City*. London: Routledge.
Kannisto, T. (2016). Positio contra complementum possibilitatis–Kant and Baumgarten on Existence. *Kant-studien*, 107(2), pp. 291–313.
Kant, I. (1998[1781]). *Critique of Pure Reason*. Translated by P. Guyer and A.W. Wood. Cambridge: Cambridge University Press.
Kant, I. (2000[1790]). *Critique of the Power of Judgment*. Translated by P. Guyer and E. Matthews. Cambridge: Cambridge University Press.
Katz, C. (1996). Towards minor theory. *Environment and Planning D: Society and Space*, 14(4), pp. 487–499.
Kompridis, N. (2014). *The Aesthetic Turn in Political Thought*. London: Bloomsbury.
Latham, A. (2009). Flâneur, The. In *International Encyclopedia of Human Geography*, edited by A. Kobayashi. Amsterdam: Elsevier. pp. 153–157.
Latham, A. and Layton, J. (2020). Kinaesthetic cities: Studying the worlds of amateur sports and fitness in contemporary urban environments. *Progress in Human Geography*, 44(5), pp. 852–876.
Lees, L. (2004). Urban geography: Discourse analysis and urban research. *Progress in Human Geography*, 28(1), pp. 101–107.
Lefebvre, H. (1970). *The Urban Revolution*. Translated by R. Bononno. University of Minnesota Press.
Leszczynski, A. and Kong, V. (2023). Walking (with) the platform: Bikesharing and the aesthetics of gentrification in Vancouver. *Urban Geography*, 44(4), pp. 773–795.
Ley, D. (2003). Artists, aestheticisation and the field of gentrification. *Urban Studies*, 40(12), pp. 2527–2544.
Macrorie, R., Marvin, S. and While, A. (2021). Robotics and automation in the city: A research agenda. *Urban Geography*, 42(2), pp. 197–217.

Manalansan, M. (2006). Immigrant lives and the politics of olfaction in the global city. In *The Smell Culture Reader*, edited by J. Drobnick. Oxford and New York: Berg, pp. 41–52.
Marchart, O. (2019). *Conflictual Aesthetics: Artistic Activism and the Public Sphere*. London: Sternberg Press.
Marchart, O. (2007). *Post-foundational Political Thought: Political Difference in Nancy, Lefort, Badiou and Laclau*. Edinburgh: Edinburgh University Press.
Marston, S. A., Jones, J. P. and Woodward, K. (2005). Human geography without scale. *Transactions of the Institute of British Geographers* 30(4), pp. 416–432.
Massey, D. (1991). Flexible sexism. *Environment and Planning D: Society and Space*, 9(1), pp. 31–57.
McFarlane, C. (2011). The city as assemblage: Dwelling and urban space. *Environment and Planning D: Society and Space*, 29(4), pp. 649–671.
McQuire, S. (2008). *The Media City: Media, Architecture and Urban Space*. London: SAGE Publications.
Millie, A. (2017). Urban interventionism as a challenge to aesthetic order: Towards an aesthetic criminology. *Crime, Media, Culture*, 13(1), pp. 3–20.
Montefrio, M.J.F., Lee, X.R. and Lim, E. (2021). Aesthetic politics and community gardens in Singapore. *Urban Geography*, 42(10), pp. 1459–1479.
Murakami Wood, D. and Abe, K. (2011). The aesthetics of control: Mega events and transformations in Japanese urban order. *Urban Studies*, 48(15), pp. 3241–3257.
Nassar, A. (2021). Geopoetics as disruptive aesthetics: Vignettes from Cairo. *GeoHumanities*, 7(2), pp. 455–463.
Nuzzo, A. (2006). Kant and Herder on Baumgarten's Aesthetica. *Journal of the History of Philosophy*, 44(4), pp. 577–597.
Pain, R. (2001). Gender, race, age and fear in the city. *Urban Studies*, 38(5–6), pp. 899–913.
Parker, S. (2000). Tales of the city: Situating urban discourse in place and time. *City*, 4(2), pp. 233–246.
Parnell, S. and Robinson, J. (2012). (Re)theorizing cities from the Global South: Looking beyond neoliberalism. *Urban Geography*, 33(4), pp. 593–617.
Pinder, D. (2013). *Visions of the City: Utopianism, Power and Politics in Twentieth Century Urbanism*. London: Routledge.
Pohl, L. (2022). Aura of decay: Fetishising ruins with Benjamin and Lacan. *Transactions of the Institute of British Geographers*, 47(1), pp. 153–166.
Pow, C.P. (2018). Building a harmonious society through greening: Ecological civilization and aesthetic governmentality in China. *Annals of the American Association of Geographers*, 108(3), pp. 864–883.
Pyyry, N. (2019). From psychogeography to hanging-out-knowing: Situationist dérive in nonrepresentational urban research. *Area*, 51(2), pp. 315–323.
Rancière, J. (1999). *Disagreement: Politics and Philosophy*. Minnesota: University of Minnesota Press.
Reiss, H. (1994). The 'Naturalization' of the term 'Ästhetik' in eighteenth-century German: Alexander Gottlieb Baumgarten and his impact. *The Modern Language Review*, 89(3), pp. 645–658.
Rhys-Taylor, A. (2013). Disgust and distinction: The case of the jellied eel. *The Sociological Review*, 61(2), pp. 227–246.
Robinson J. (2011). Cities in a world of cities: The comparative gesture. *International Journal of Urban and Regional Research*, 35(1): 1–23.
Roy A. (2009). The 21st-century metropolis: New geographies of theory. *Regional Studies*, 43(6): 819–830.
Sassen, S. (1991). *The Global City: New York, London, Tokyo*, Princeton, NJ: Princeton Univ.

Scott, A.J. and Storper, M. (2015). The nature of cities: The scope and limits of urban theory. *International Journal of Urban and Regional Research*, 39(1), pp. 1–15.
Shaviro (2007). The 'Wrenching Duality' of Aesthetics: Kant, Deleuze, and the 'Theory of the Sensible.' Accessed June 01 2024. Available from: http://www.shaviro.com/Othertexts/articles.html
Sheppard E, Leitner H and Malignant A (2013). Urban pulse – Provincializing global urban-ism: A manifesto. *Urban Geography*, 34(7), pp. 893–900.
Shusterman, R. (1999). Somaesthetics: A disciplinary proposal. *The Journal of Aesthetics and Art Criticism*, 57(3), pp. 299–313.
Simmel, G. (1950[1903]). The Metropolis and Mental Life. In *The Sociology of Georg Simmel*, edited by K.H. Wolff. New York: Free Press, pp. 409–424.
Simone, A. (2016). Urbanity and generic blackness. *Theory, Culture and Society*, 33(7–8), 183–203.
Simone, A. (2020). Cities of the Global South. *Annual Review of Sociology*, 46(1), pp. 603–622.
Sjöholm, C. (2015). *Doing Aesthetics with Arendt: How to See Things*. Columbia: Columbia University Press.
Swyngedouw, E. (2009). The antinomies of the postpolitical city: In search of a democratic politics of environmental production. *International Journal of Urban and Regional Research*, 33(3), pp. 601–620.
Taylor, N. (2003). The aesthetic experience of traffic in the modern city. *Urban Studies*, 40(8), pp. 1609–1625.
Tedesco, D. and Davies, M. (2022). Cities as aesthetic subjects. *Globalizations*, pp. 1–17. Available at: https://doi.org/10.1080/14747731.2022.2117505 (Accessed 1st January 2025).
Tursić, M. (2019). The city as an aesthetic space. *City*, 23(2), pp. 205–221.
UN Habitat (2022). World Cities Report 2022 [online]. Available from: https://unhabitat.org/sites/default/files/2022/06/wcr_2022.pdf
Usher, M. (2023). Design and revolution: Morris, modernism and urban gothic. *Urban Geography*, 44(6), pp. 1117–1145.
Virmani, A. (ed.) (2021). *Aesthetic Perceptions of Urban Environments*. London: Routledge.
Walker, R. (2015). Building a better theory of the urban: A response to 'Towards a new epistemology of the urban?'. *City*, 19(2–3), pp. 183–191.
Ward, K. (2008). Toward a comparative (re) turn in urban studies? Some reflections. *Urban Geography*, 29(5), pp. 405–410.
Watt, P. (2006). Respectability, roughness and 'race': Neighbourhood place images and the making of working-class social distinctions in London. *International Journal of Urban and Regional Research*, 30(4), pp. 776–797.
Wilson, H.F. (2017). On geography and encounter: Bodies, borders, and difference. *Progress in Human Geography*, 41(4), pp. 451–471.
Wilson, H.F. (2016). Encountering Havana: Texts, Aesthetics and Documentary Encounters. In: *Encountering the City: Urban Encounters from Accra to New York*, edited by J. Darling and H. Wilson. London: Routledge. Ch.12.
Yusoff, K. (2010). Biopolitical economies and the political aesthetics of climate change. *Theory, Culture & Society*, 27(2–3), pp. 73–99.
Yusoff, K. (2015). Geologic subjects: Nonhuman origins, geomorphic aesthetics and the art of becoming in human. *Cultural Geographies*, 22(3), pp. 383–407.

2 Perspectives from Philosophical and Applied Urban Aesthetics

Sanna Lehtinen

Introduction

Cities have been studied more concertedly over the past 50 years from a variety of disciplinary perspectives with an interest in the aesthetic qualities and stimuli they offer to the human senses.[1] What has been common to these approaches is to consider the city as a place of interaction between humans and the material composition of their environment. Mere sense perception is not, however, at the core of the specifically philosophical interest in the aesthetics of cities. Rather, one should also consider the evaluations and judgements that are *based upon* sense perception. Aesthetic experience is, then, ultimately linked to broader and socially contested aesthetic values which have longer-term significance for how cities are designed, developed, and used. Aesthetic values are difficult to home in on, as they have a high degree of subjectivity even when shared by people in the same community. With the help of philosophical urban aesthetics, the value formation processes can nonetheless be studied, even in cases when values are debated or conflicting.

What differentiates the various newer approaches from traditional architectural theory that shows an interest in the aesthetic quality of cities is the focus on the urban environment beyond the clearly designed and planned elements of the city, taking into better consideration the unplanned and unintentional results of both human and nonhuman activity. Although built and designed spaces are central to the make-up of any urban environment, unplanned elements greatly contribute to how a city is perceived and experienced by residents and visitors alike. Taking various elements of cities into consideration, including their change and development and relations and intangible factors, demands an even more discerning and relational approach from the contemporary study of urban aesthetics.

What is special about the philosophical perspective to urban aesthetics is (i) the interest in the processual nature of aesthetic experiences, (ii) their respective components, and (iii) how collectively contested aesthetic values are linked to the subjective facets of these experiences. In addition to an interest in sense perception, philosophical urban aesthetics focuses on the role of human cognitive capacities in shaping (and being shaped by) aesthetic perception and

DOI: 10.4324/9781003335993-2

preference. This refers to how our knowledge, beliefs, and insights about whatever it is that we are perceiving affects our aesthetic judgement. In philosophical aesthetics and aesthetic theory, since Immanuel Kant, there is a strong tradition which emphasises the "disinterestedness" of aesthetic judgements. In other words, it has been claimed that in order for an aesthetic judgement to truly concern the aesthetic dimension, it has to be detached from other possible interests towards the object under assessment (Zangwill, 2023).

This assumption has, however, been severely questioned, as the objects of aesthetic interest and systematic study have become both broader and more complex in recent decades. The broadening refers simultaneously to the opening up of the academic field towards new directions such as everyday and environmental aesthetics, but also to the overall processes of aestheticisation which testify of almost any element of human life having an aesthetically relevant dimension (Welsch, 1996). It is now generally assumed that a more interesting and engaged object of thought is everyday aesthetic interactions. This emphasis on the everyday is closer to the original definition of aesthetics by Alexander Gottlieb Baumgarten in the 18th century where in *Aesthetica* (1750) he termed aesthetics as "the science of sensible knowing" or perfection of sensuous cognition. The definition does not delimit the field of aesthetics to concern only the sphere of art or human creativity in general but frames it as a manner of gaining information and furthering knowledge about all perceivable facets of the world. Art is not, then, the centre of attention nor is beauty the only aesthetic quality of interest.

Moreover, more recently, emphasis has been placed on the relevance of everyday life as a context for experiencing the city. Especially after the millennium, the diversity of everyday aesthetic phenomena and "everydayness" itself has been of increasing attention. What the everyday means in the context of everyday aesthetics is not easy to define, but it can either be grasped through the practices and features of everyday life such as "food, wardrobe, dwelling, conviviality, and going out" (Melchionne, 2013: 5) or through normal, habitual, and routine "*objects, activities, and events*, as well as certain *attitudes and relations* to them" (Naukkarinen, 2013: 3, emphasis original). The processual nature of the everyday is anchored by the perceivable objects, phenomena, organisms, and structures of the city. In this sense, the phenomena of relevance to aesthetics can include the daily commute, access to nature, weather conditions and elements of technological infrastructure. In the light of foundational research on everyday aesthetics, the range of aesthetically relevant phenomena is as vast as the scope of our senses and the span of our attention (Light and Smith, 2005; Saito, 2007; Leddy, 2012).

In this chapter, I discuss some of the main components of the aesthetic dimension of cities through and with the framework of philosophical aesthetics. Around the latter, a distinction might be made between aesthetic theory and more applied forms of urban aesthetics, but it is preferable to think of the application of philosophical insight as happening on a sliding scale. All aesthetic knowledge is applied in the sense that it evolves and becomes acknowledged in the context of lived and engaged experiences with ever-changing environments. Cities form a very particular case for aesthetic experience and appreciation, and traditional forms of aesthetic theory need to be applied accordingly and with sensitivity to changes in context.

Another interesting area of insight is the more recent empirical theories on urban aesthetics which this chapter only has scope to briefly mention. Such work, focusing on the psychological impact of aesthetic experiences, is currently being developed beyond the previous paradigm of psychology-linked theories on environmental preferences and will no doubt have an important contribution to the development of philosophical urban aesthetics (see Besson, 2019). The main task of this chapter, however, is to present and further develop philosophical aesthetic theory in a way that acknowledges and meets the existing conditions of contemporary cities globally.

Perspectives from Philosophical and Applied Urban Aesthetics

Urban aesthetics is central to the study of cities as sensescapes or as otherwise experientially central and vibrant areas of human activity. In recent years, concepts such as livability have been applied to the study of the experiential aspects of urban life (Stefansdottir, 2014). Intuitively, urban environments seem to offer incessant sensory stimuli for every taste and second of the day. How we then process and further assess these stimuli is much more difficult to trace. Cities entail an assortment of environmental, art-related, and everyday factors. The combined knowledge of the three main theoretical pillars of contemporary philosophical aesthetics can thus be applied to cities through the field of urban aesthetics. The term "urban aesthetics" is relatively new. It has been used in philosophical and applied aesthetics since the 1990s, although it has been present in some branches of urban planning already in the 1950s (Haapala, 1998; Williams, 1954). However, in the planning-focused literature, the understanding of aesthetics is quite thin compared to how this dimension of human experience is approached in environmental aesthetics, for example.

Urban aesthetics, understood in this sense as a branch of philosophical aesthetics, is also an important part of a relatively recent philosophical orientation called the philosophy of the city (PotC) which brings together many different facets of the philosophical discussion with an interest in contemporary cities globally. In the context of the PotC, it is possible to focus on ontological, ethical, epistemological, or aesthetic questions in relation to concrete urban issues and this is usually made by bringing together several theoretical and practical approaches to better serve the inquiry into contemporary urban environments (c.f. Epting, 2018; Meagher and Noll, 2019; Lehtinen, 2020a; Simon, 2021; Lehtinen and Lobo, 2022).[2] In many philosophical questions concerning the city, the appearance of elements of the urban environment is central. This is mainly because the appearance and perceivable features are linked to their ability to draw attention in the first place, but also to how the appearance of things can make other values of the community perceivable.

The background of more contemporary theories on philosophical urban aesthetics is largely owed to developments made in environmental aesthetics in the past few decades. The current research paradigm of environmental aesthetics started to gain traction as a distinct area of philosophical aesthetics in the early 1960s. The origin of environmental aesthetics can be traced to the article "Contemporary Aesthetics and the Neglect of Natural Beauty" by Ronald Hepburn (1966). The perceived and already alarming anthropogenic change of natural environments and rising environmental worry caused by its most visibly destructive forms was behind the surge in interest in the aesthetic aspects of natural environments. When considering the development of environmental aesthetics to the specific domain of urban aesthetics, it is important to keep in mind this normative starting point, deeply rooted in environmental concerns. The development of environmental aesthetics has significantly contributed to shaping and advancing philosophical urban aesthetics, particularly within contemporary sustainability science and studies, where urban challenges play a critical role.

An example of where environmental aesthetics is shaping and advancing philosophical urban aesthetics is through a focus on aesthetic disillusionment. Aesthetic disillusionment refers to losing one's sense of pleasure or emotional connection in relation to something previously perceived as beautiful or otherwise meaningful. It is a phenomenon that depicts the experience of friction created by the relation between our understanding and knowledge of an event and our aesthetic appreciation (Foster, 1992). When epistemic beliefs change, the perceptual elements of the experience do not change: the surface qualities of the object of appreciation remain the same, but something in our relation to it changes and this causes changes in the experience as well. An example of this kind of aesthetic disillusionment is if we have admired a red sunset over the city, but then gain knowledge that it has been caused by air pollution in the lower layers of the atmosphere. We might still consider the view beautiful but this new knowledge of the environmentally detrimental phenomenon that is causing the beautiful red hues nonetheless changes our reaction. The change in the aesthetic judgement might be rapid and change completely, but it is often a process that takes time after the initial friction has been put in motion. This is an especially valuable insight for the study of sustainable consumption, where the hope is that ecological awareness would also change aesthetic taste to prefer more sustainable options (Lehtinen, 2021b). In cities, this phenomenon is also visible in the renewed appreciation of architectural styles that have been considered inferior at some point. The revival of interest in Brutalist architecture—once criticised for its stark, utilitarian design—from the 1970s is a recognised example.

Cityscape: The Macro Layer of Urban Aesthetics

As aesthetic issues are omnipresent in the everyday dimension of urban life, drawing further distinction between "landscape aesthetics" and "everyday aesthetics" is useful. One way to approach this distinction is through the notions of the macro and micro layers of urban aesthetics (Lehtinen, 2020b). While this distinction serves as a theoretical tool, it reflects the breadth and diversity of urban aesthetic experiences in practice. The development of terminology around urban aesthetics should aim to both describe and validate these experiences, as seen in how broadly the term "aesthetics" is used and understood in everyday conversations and the media. Categorising these experiences by scope or scale (i.e. "the macro" or "the micro") is a methodological tool, intended to aid understanding rather than provide a strict representation of reality.

The so-called *macro aesthetic layer* describes a distinctly distanced view of a city, focusing on how it is perceived as a "whole" or from a broad vantage point. The macro aesthetic layer can be present in the context of a particular city but also generalised to consider how cities more broadly are imagined. The macro layer is present when discussing urban development through images and computer-generated visualisations, for example. It is also the focus of discussion on phenomena such as city branding or loaded preconceptions regarding how certain neighbourhoods look. The macro aesthetic approach is visually dominated, although it also takes other senses into consideration. Real-world, empirical engagement with the environment ranges from minimal (the city perceived through images, visualisations, or mental reproductions) to distanced (admiring the cityscape from a distance, through the lens of a camera or smartphone, or from a moving vehicle). In environmental aesthetics, this type of minimal engagement is discussed as landscape aesthetics or a landscape model of aesthetic appreciation stemming 'directly from the tradition of the picturesque' and 'framing and flattening environments into scenery' (Carlson, 2023: np). The landscape model is challenged by some philosophers such as Arnold Berleant with more engaged and "on the ground" modes of appreciating surroundings (Berleant and Carlson, 2007; Berleant, 2010).

How cities appear at the cityscape or macro level is always a compromise: we live with the remnants of the decisions of previous generations (Lehtinen, 2022). This applies especially to historic cities in which their historical trajectories have made them the visually recognisable places that they are. Emblematic cities such as Paris, New York, or Tokyo, for example, are easy to recognise based on their landmark monuments and prevalent architectural styles. Furthermore, the ongoing planning and design activities have to take this into consideration as there is less space for new development in the dense centres of many historically rich cities. Experientially, this type of strong aesthetic identity must also be considered in relation to the diverse, coexisting generations of citizens, each of whom may hold vastly different views on how the city should or could ideally appear. In recent research in philosophical aesthetics, this is described through the concepts of inter- and transgenerational aesthetics, which show that aesthetic choices and preferences are not produced in a vacuum, independently of the choices and preferences of others, often spanning vast spatial and temporal distances (Lehtinen, 2020a; Capdevila-Werning and Lehtinen, 2021a; Capdevila-Werning and Lehtinen, 2021b).

The so-called macro dimension of aesthetic experience is less dependent than the micro-aesthetic layer on the manner and degree of bodily engagement.

However, it can also rely on multi-sensory aesthetic qualities such as smell and the tactile features of an urban environment. The macro layer refers to more distanced and conceptualised ideas about how a city looks or feels. To some degree, it does not necessarily even need the city to be there to be perceived: it is enough that its features and easily recognisable elements are imagined, remembered, or represented through other means. This opens a path for nostalgic interpretations of how a city should or could look in the normative sense: traditionalist and conservative views in urban planning or architecture, for example, potentially feed and flourish on the macro layer where the past appearance of places is imagined to be more glorious than it might have been in reality (see, for instance, Gassner, 2019; Lohtaja, 2022). On the other hand, more progressive utopian visions also feed from this more detached way of contemplating the city—seeing the city from a distance, either literally or conceptually. However, this abstract, distanced or idealised way of thinking about the city is quite different from the immediate, direct experience of a city's aesthetics where the presence of the object of appreciation (the city—and all of its sights, sounds, and tangible qualities) is considered to be of central importance (Carlson, 2023).

Aesthetic Undertone of Everyday Urban Life: The Micro Aesthetic Layer

In comparison to the macro aesthetic layer, the *micro aesthetic layer* describes a more bottom-up and hands-on engagement with the aesthetic features of an urban environment. The micro layer takes into consideration the insights from everyday aesthetics as it describes the fluctuation between the familiar and the strange elements affecting the urban aesthetic experience (Haapala, 2005). An example of this is the use and reliance on routinised walking routes in one's own neighbourhood: mostly the familiar corners stay the same but there are always unexpected things to experience as well, adding an element of strangeness and surprise even to the most habitual daily walk. The fluctuation between a type of pre-reflexive familiarity and aesthetic interest towards new elements is thus a basic condition that shapes which facets of the environment receive more sustained attention. In familiar environments, routines and habits can affect how environments are observed. The senses are operating in a similar way as before but the so-called extraordinary "peak experiences" are missing because attention is not actively drawn or directed to them (Leddy, 2012). Art in the urban context is a paradigmatic example of a peak experience, as it invites to stop and pay attention to and through its perceptual features.

The possible numbing in relation to the most mundane and familiar features can be contested by the purposeful use of means such as digital tools, such as a camera, to draw attention back to the features of places that would otherwise go unnoticed (Lehtinen and Vihanninjoki, 2021). There are factors which might occasionally draw one to view even the most familiar environment in a new light. New digital technologies or different tactics of gamifying the urban experience, for example, increasingly encourage new ways of seeing. GPS-based navigation is one relatively recent yet already ubiquitous example of this. When engaging in playing games such as Pokémon GO, the details such as architectural ornaments in the most familiar surroundings gain new importance. Equally, the use of route planners can mean that a person's focus is freed so they can observe surroundings instead of having to concentrate on their exact location on the map (Lehtinen and Vihanninjoki, 2019).

The aesthetic undertone of everyday life becomes recognised through the micro aesthetic layer (Lehtinen, 2020b). This refers to the habituation that takes place when we encounter something on a daily basis in our daily environment. The accumulation of aesthetic factors can be said to have an impact even when it does not cause peak aesthetic experiences. The type of environment in which one spends the majority of their everyday life is therefore highly significant. Aesthetic factors play an important role in envisioning, creating, and cultivating the lived quality or livability in urban environments. Phenomenological traditions of thinking about human-world relations with its emphasis on multi-sensory bodily presence are influential to all developments in philosophical urban aesthetics, particularly for developments surrounding the macro aesthetic layer. The idea of aesthetic engagement developed by Berleant takes insight from the phenomenological tradition and brings it together with a pragmatist, John Dewey-inspired understanding of an especially engaged form of experience (Berleant, 2010; Berleant, 2012).

As well as focusing on the fluctuation between the new and the familiar, the micro aesthetic layer is characterised by the accumulation of minor and habitual impressions. These are formed through the seemingly insignificant moments of aesthetic pleasure or displeasure. Exposure to the more mundane layers that make the city might not be recognised as the main highlight of one's aesthetic life, but they nonetheless build the foundation for extraordinary experiences too (Leddy, 2012). By comparison, the micro aesthetic dimension refers to the moments when the city is there to be perceived, even in close proximity. The micro aesthetic layer describes a more embodied multi-sensory experience when touch, smell, auditory, and other sensory elements merge with the visual perception of what is to be encountered. This type of aesthetic engagement is also less determined by an intention of reaching a conclusion in aesthetic judgement: the ensuing experience does not have to be either beautiful or ugly; to use paradigmatic aesthetic terms, it can be more ambivalent in nature. Naturally, in real-life situations, the representation-oriented macro and the lived experience dimension of the microlayer often merge. It is possible to zoom in and out in situations in which aesthetic engagement and apprehension take place. This type of oscillation between taking a closer look and gaining insight into the bigger more contextual picture is to be recognised as part of the aesthetic appreciation of the environment.

In addition to the macro and micro layers, a third way of approaching the city aesthetically can be distinguished. This perspective focuses on the experience of urban complexity, which is closely aligned with traditional accounts of *the sublime* (Lehtinen, Strandhagen, and Tainio, 2023). The sublime refers here to the awe-inspiring recognition that one is part of something much bigger than oneself. By perceiving part of the city, we are able to grasp only some facets of the processual nature of the city as an organism. This describes the experience of being part of the invisible—yet clearly existing—structures of the city that hold it together and make it a functioning entity. The particularly *urban* sublime has been described in literary studies through the notion of the "technological sublime"[3] and also in relation to the anonymous masses of diverse urbanites in the city, "the multitudinous humanity" (Den Tandt, 1994; Den Tandt, 2014: 126). The previous descriptions of the urban sublime stem, however, mainly from literary descriptions of cities, thus being at least one step distanced from the actual engaged experiences of the city which philosophical urban aesthetics is mainly focused on. The literary descriptions are still helpful in conceptualising these experiences but cannot serve as the only source for analysing sublimity in the context of urban aesthetics. An example of urban sublimity is the transportation system: the actual metro grid of any global metropolis is impossible to grasp perceptually, yet we know it exists and rely on it as an important element of everyday life. The sublime in the urban context can also be compared to the sublime feelings that natural environments evoke. For example, when the imposing cityscape of Manhattan is compared to the magnificent silhouette of the Rocky Mountains (Vogel, 2015).

Art in the Urban Context

One way to exemplify the theoretical distinction between the macro and micro layers of urban aesthetics is through public art or other forms of artistic activity in the city. Art in the urban context is an increasingly broad category of different types of artistic and otherwise creative human outputs that take form and place in the urban environment. The city provides an exceptionally rich and diverse environment for artistic activity as it is, itself, a forever-changing platform. Cities also differ in this respect. Depending on their qualities, different types of art are seen as fitting better in different types of cities, or areas within them. For example, a pedestrian-friendly environment affords artworks that take a longer time to observe whereas more car-dependent areas afford works that are observable from a moving vehicle. This is reflected in the size, composition, and content of the artworks.

With the aesthetic dimension of urban art in focus, it is also noteworthy that not all urban or public art is recognised as being qualitatively good or experientially interesting for all relevant audiences or perspectives (Kingwell, 2019). Nor is all urban or public art necessarily available for predominantly aesthetic modes of evaluation: art in the urban context can also be highly conceptual without a clear sensory presence in the cityscape, or the sensory aspect might be secondary or ephemeral.[4] Even then, however, urban art most likely has consequences for how the city is perceived. Art can do this by nudging its audience towards new interpretations of the existing elements of the city with its perceptually available means. Artworks that use the street space over the flow of traffic, for example, quite literally direct the gaze and attention to parts of the streetscape and buildings that one may not ordinarily pay attention to. It does this, firstly, by attracting attention with its perceptual cues followed by a process of sustained aesthetic interest in the intricate meanings it conveys (Lehtinen, 2015).

Aesthetics as a discipline has, for a long time, been associated with philosophical and theoretical questions specifically in the context of art. Such an interest has broadened significantly over the past half a century. Nonetheless, artistic phenomena still act as an important starting point for reflection and an example of many aesthetically central issues and questions in the urban environment. Urban art is a broader category than public art, taking into consideration the bottom-up artistic movements and activities that do not fall solely under the category of art, such as craftivism or guerilla gardening, that use urban space in creative ways. Aesthetic theory can still be applied to them as this type of grassroots activity affects the perceptual qualities of the urban environment by staking a claim in a particular environment. Community-led creative activity also engages people and provides a sense of active agency in their own everyday environments. This most likely increases aesthetic satisfaction and overall well-being over time. What is more, art can raise difficult questions and make visible possible points of tension and bias in the urban context. This is not the duty of art, sometimes its function can be about sensory pleasure without normative or ameliorative dimension. Urban environments, then, are not purely functional. Rather, they are places in which human imagination can flourish and where play is used to engage with surroundings.

Further Directions in Philosophical Urban Aesthetics

Everyday aesthetics has perhaps been the single most influential factor in developing philosophical urban aesthetics since the turn of the 21st century. Acknowledging that everydayness is at the core of urban experience facilitates taking the city into consideration in a radically different sense than with the previous paradigm in aesthetic theory which tended to focus on extraordinary experiences. The ordinary forms a basis for the extraordinary to a great extent, but the routine, habitual exchanges form a particular level of engagement. Everyday small engagements accumulate into ways of seeing and paying attention which, in turn, significantly mould aesthetic sensibilities (Korpelainen, 2021). This process takes place in both directions whereby aesthetic sensibility also increases awareness of more ordinary, quotidian aesthetics. What one pays attention to in the first place is thus not an accident but is formed through and with daily interaction and the affordances of familiar environments.

Philosophical aesthetics sheds light on how urban environments become perceived and appreciated. It is, however, clear that the studying of urban sensescapes is and should be a transdisciplinary endeavour. Further study of macro and micro aesthetic layers would benefit from the richness of the various different theoretical and disciplinary vantage points that a transdisciplinary approach facilitates Moreover, these efforts in research on urban aesthetics need to be closely aligned to what is actually happening in and around cities: theoretical knowledge needs to be accompanied by, and closely aligned with, practical knowledge of designing and observing urban environments in their everyday use.

Urban technologies and their maintenance present an interesting area for the study of urban aesthetics. However, defining these technologies can be challenging, which the PotC has recently started to consider (Nagenborg et al., 2021). What impacts urban technologies will have on the aesthetic dimension extends this inquiry, considering the perceptual and aesthetic effects of implementing and using various technologies (Lehtinen, 2021a). From the perspective of aesthetics, those technologies that have some direct or indirect aesthetic effects are of further interest. Contemporary telecommunications technologies, for example, are of specific macro aesthetic interest when it comes to the physical structures such as antennas and cell towers which are required to ensure network coverage. In addition, the implementation of 5G and the subsequent 6G network has other effects on the use of the city: their ultra-low latency should enable high-quality video streaming and the use of more complex Augmented Reality (AR) or Virtual Reality (VR) solutions (Lehtinen and Fantini Van Ditmar, 2022). This means that 5G and 6G will have an indirect effect on the micro urban aesthetic dimension too. AR solutions that are used for architectural reconstruction are an example of this (Wittinsglow and Lehtinen, 2024).

One of the main goals of this chapter has been to demonstrate that philosophical knowledge about urban aesthetics can be of help in defining central concepts such as beauty or aesthetic value. It also encourages critical inquiry into what roles artistic activity or creativity in general play in the urban context, while speculatively proposing and examining possible future scenarios for comprehending the city. Aesthetic values are not independent of other values; they are rather born and developed (to a great extent) as a reflection of what is otherwise considered to be valuable in society. This societal deliberation is shaped by the multitude of experiences of individual urbanites, which form in response and in relation to the shared urban environment, including its material and immaterial stimuli.

How to reassess prevailing aesthetic judgements becomes increasingly central as cities are planned and developed towards the future as they shape the environments in which we live. This reassessment is essential to ensure that cities are inclusive and supportive of human and non-human life. This is a task for humanity, and in this regard, human-centred forms of aesthetic theory are valuable for understanding what we prefer and value, why this is the case, and whether and how this can be changed to improve living conditions in cities for other species as well. A multi-agency understanding of aesthetics, though not mainstream in traditional philosophical aesthetic theory, is becoming increasingly central for evaluating the practical consequences of aesthetic preferences and values. This need for developing new normative approaches within philosophical aesthetics stems from the recognition that ethical conditions always frame aesthetic experiences as they take place. While traditional concepts of disinterestedness are important in certain aesthetic contexts, they are increasingly challenged when aesthetic theory is applied to everyday or diverse human environments.

Conclusions

This chapter has explored the perspective of philosophical and applied urban aesthetics in the study of contemporary urban environments. Its aim has been to present this approach whilst contributing to defining what aesthetics means in the urban context, with consideration of the category of the "city", and how its role for the human experience can be analysed and assessed. By tracing some central and prevalent lines of thought from aesthetic theory through environmental aesthetics, the discussion has highlighted its key tenets and distinctions, such as between the macro and micro layers of urban aesthetics. The "macro" layer relates to those imagined, represented, and remembered elements and the "micro" layer addresses direct, sensory encounters with the city and urban phenomena more generally. These layers help to illuminate the different ways of experiencing and discussing the aesthetic dimension of cities, for example.

In addition, the traditional concept of the sublime—the awe-inspiring recognition that one is part of something much bigger than oneself—has been discussed in the urban context. So too has the concept of aesthetic disillusionment, how our judgement that something is (not) beautiful or significant (such as formerly revered or reviled architectural styles that come to seemingly define cities) are altered through shifts in social and cultural values. Both have been presented as a useful tool for considering complex experiences in contemporary cities.

In this volume, aesthetics is used as a term to describe how we perceive and make sense of the city. I have made the case that philosophical and applied urban aesthetics offer some central conceptual tools for this endeavour and called for interdisciplinary urban scholars to make use of its conceptual tools to this end. For further understanding multi-sensory experiences, this chapter has homed in on elements such as art and technology in the urban context. While urban experiences often involve a blend of elements (values, encounters, perceptions, and so forth), the conceptual framework of philosophical and applied urban aesthetics allows us to zoom in on specific facets, offering insight into what constitutes these experiences.

Furthermore, as a next step for philosophical and applied urban aesthetics, it is crucial to recognise that gaining aesthetic insight into contemporary cities worldwide is a distinctly inter-, multi-, and transdisciplinary endeavour. This point is also emphasised by one of the main motivations of this edited volume: to ask what it means to think through an aesthetics of the city. Such an endeavour requires bridging the sensory and the cognitive, linking perceptual questions to broader questions around values, and we ought to not consider them in isolation. It also goes to show that beauty is not the only concept of interest in aesthetics (a matter of values) nor is it only in the eye of the beholder (a matter of perception). Attending to the complexity of both sides of this coin simultaneously to make sense of the city can only be benefitted by working beyond our disciplinary siloes.

Notes

1 See Lefebvre (1974) on the production of space or Madsen and Plunz (2002).
2 For info on Philosophy of the City Research Group, see https://www.potcrg.org.
3 A term coined first by Perry Miller, see Nye (1994).
4 As an example of this the highly debated yet influential Olafur Eliasson's *Ice Watch* (2014), https://olafureliasson.net/artwork/ice-watch-2014.

References

Berleant, A. (2010). *Sensibility and Sense: The Aesthetic Transformation of the Human World*. Exeter: Imprint Academic.

Berleant, A. (2012). *Aesthetics Beyond the Arts: New and Recent Essays*. Farnham & Burlington, VT: Ashgate.

Berleant, A. and Carlson, A. (eds.) (2007). *The Aesthetics of Human Environments*. Peterborough. ON: Broadview Press.

Besson, A. (2019). "In defense of cities: On negative presentation of urban areas in environmental preference studies", *Contemporary Aesthetics*, 17, Article 861. https://contempaesthetics.org/newvolume/pages/article.php?articleID=861

Capdevila-Werning, R. and Lehtinen, S. (2021a). "Intergenerational aesthetics: A future-oriented approach to aesthetic theory and practice", *Philosophical Inquiries*, 9(2) (special issue "Philosophy of Art. New Directions", ed. E. Caldarola and J. Levinson), pp. 175–198.

Capdevila-Werning, R. and Lehtinen, S. (2021b). "A first approach to intergenerational aesthetics: Theoretical stakes, practical examples, and future research avenues", *European Society for Aesthetics ESA Conference Proceedings*, 13, pp. 43–58.

Carlson, A. (2023). "Environmental Aesthetics", In *The Stanford Encyclopedia of Philosophy (Fall 2023 Edition)*, edited by Edward N. Zalta and Uri Nodelman, https://plato.stanford.edu/archives/fall2023/entries/environmental-aesthetics

Den Tandt, C. (1994). *The Urban Sublime in American Literary Naturalism*. Champaign, IL: University of Illinois Press.

Den Tandt, C. (2014). "Masses, forces, and the urban sublime", In *The Cambridge Companion to the City in Literature*, edited by Kevin R. McNamara. Cambridge: Cambridge University Press, pp. 126–136.

Epting, S. (2018). "Introduction to the Special Issue on Philosophy of the city and environmental ethics", *Environmental Ethics*, 40(2), pp. 2–3.

Foster, C. (1992). "Aesthetic disillusionment: environment, ethics, art", *Environmental Values*, 1(3), pp. 205–215.

Gassner, G. (2019). *Ruined Skylines: Aesthetics, Politics and London's Towering Cityscape*. London: Routledge.

Haapala, A. (2005). "On the Aesthetics of the Everyday. Familiarity, Strangeness, and the Meaning of Place," In *The Aesthetics of Everyday Life*, edited by Andrew Light and Jonathan M. Smith. New York: Columbia University Press, pp. 39–55.

Haapala, A. (ed.) (1998). *The City as Cultural Metaphor: Studies in Urban Aesthetics*. Lahti: International Institute of Applied Aesthetics.

Hepburn, R. (1966). "Contemporary Aesthetics and the Neglect of Natural Beauty", In British Analytical Philosophy, edited by B. Williams and A. Montefiore. London: Routledge and Kegan Paul.

Kingwell, M. (2019). "Editorial Introduction to the Topical Issue 'Does public art have to be bad art?'", *Open Philosophy*, 2, 1. https://doi.org/10.1515/opphil-2019-0041
Korpelainen, N. (2021). "Cultivating aesthetic sensibility for sustainability", *ESPES. The Slovak Journal of Aesthetics*, 10(2), pp. 165–182.
Leddy, T. (2012). *The Extraordinary in the Ordinary: The Aesthetics of Everyday Life*. Peterborough, ON: Broadview Press.
Lefebvre, H. (1974). *La production de l'espace*. Paris: Anthropos.
Lehtinen, S. (2022). "Urban Experience as Aesthetic Compromise", In *Imperfectionist Aesthetics in Art and Everyday Life*, ed. Peter Cheyne. London: Routledge, pp. 363–374.
Lehtinen, S. (2021a). "Urban Aesthetics and Technology", In *Oxford Handbook of Philosophy of Technology*, ed. S. Vallor. Oxford: Oxford University Press, pp. 449–469.
Lehtinen, S. (2021b). "Aesthetic Sustainability", *Situating Sustainability: A Handbook of Contexts and Concepts*, edited by R. Toivanen and P. Krieg, Helsinki: Helsinki University Press, pp. 255–268.
Lehtinen, S. (2020a). "An Intergenerational Approach to Urban Futures: Introducing the Concept of Aesthetic Sustainability", In *Moving from Landscapes to Cityscapes and Back: Theoretical and Applied Approaches to Human Environments*, edited by A. Haapala, B. Frydryczak and M. Salwa. Lodz: Przypis1, pp. 111–119.
Lehtinen, S. (2020b). "Editorial introduction to the topical issue on philosophy of the city", *Open Philosophy*, 3(1), pp. 730–735.
Lehtinen, S. (2015). *Excursions into Everyday Spaces – Mapping Aesthetic Potentiality of Urban Environments through Preaesthetic Sensitivities* (doctoral dissertation, monograph), Helsinki: University of Helsinki.
Lehtinen, S., Strandhagen, B. and Tainio, M. (2023). "Rethinking the Urban Sublime", In *Applying Aesthetics to Everyday Life. Methodologies, History, and New Directions*, edited by Lisa Giombini and Adrian Kvokacka. London: Bloomsbury, 103–115.
Lehtinen, S. and Fantini Van Ditmar, D. (2022). "Introduction: The rise of 5G", *Mediapolis: A Journal of Cities and Culture*, 7(3), https://www.mediapolisjournal.com/2022/09/rise-of-5g
Lehtinen, S. and Lobo, T. (2022). " Editorial introduction to the topical issue on philosophy of the city", *East Asian Journal of Philosophy EAJP*, 1(3), pp. 1–6.
Lehtinen, S. and Vihanninjoki, V. (2021). "Aesthetic Perspectives on Urban Technologies: Conceptualizing and Evaluating the Technology-Driven Changes in the Urban Experience", In *Technology and the City: Towards a Philosophy of Urban Technologies*, edited by M. Nagenborg, M. G. Woge, T. Stone and P. Vermaas. Dordrecht: Springer, pp. 13–25.
Lehtinen, S. and Vihanninjoki, V. (2019). "Seeing new in the familiar: Intensifying Aesthetic engagement with the city through new location-based technologies", *Behaviour & Information Technology*, 39(6), 648–655. https://doi.org/10.1080/0144929X.2019.1677776.
Light, A. and Smith, J.M. (eds.) (2005). *The Aesthetics of Everyday Life*. New York: Columbia University Press.
Lohtaja, A. (2022). *Architecture as Spatial Configuration of Politics* [Doctoral dissertation]. University of Jyväskylä. JYU Dissertations, 524. http://urn.fi/URN:ISBN:978-951-39-9161-6.
Madsen, P. and Plunz, R. (2002). *The Urban Lifeworld: Formation, Perception, Representation*. London & New York: Routledge.
Meagher, S. M., Noll, S. and Biehl, J.S. (eds.) (2019). *The Routledge Handbook of Philosophy of the City*. London & New York: Routledge.
Melchionne, K. (2013). The definition of everyday aesthetics," *Contemporary Aesthetics*, 11, article 26. https://digitalcommons.risd.edu/liberalarts_contempaesthetics/vol11/iss1/26

Nagenborg, M. et al. (eds.) (2021). *Technology and the City: Towards a Philosophy of Urban Technologies*. Cham: Springer.

Naukkarinen, O. (2013). "What is 'Everyday' in everyday aesthetics?," *Contemporary Aesthetics*, 11, article 14. https://digitalcommons.risd.edu/liberalarts_contempaesthetics/vol11/iss1/14.

Nye, D. (1994). *American Technological Sublime*. Cambridge, MA: MIT Press.

Saito, Y. (2007). *Everyday Aesthetics*. Oxford: Oxford University Press.

Simon, J. (2021). "Introduction: Introducing *philosophy of the city*", *Topoi*, 40(2), 387–398. doi:https://doi.org/10.1007/s11245-021-09739-0

Stefansdottir, H. (2014). "Urban routes and commuting bicyclist's aesthetic experience", *FormAkademisk*, 7:2, Article 4, 1–21.

Vogel, S. (2015). *Thinking Like a Mall: Environmental Philosophy After the End of Nature*. Cambridge, MA: MIT Press.

Welsch, W. (1996). "Aestheticization processes: phenomena, distinctions and prospects", *Theory, Culture & Society*, 13(1), 1–24. https://doi.org/10.1177/026327696013001001

Williams, S.H. (1954). "Urban aesthetics: An approach to the study of the aesthetic characteristics of cities," *The Town Planning Review*, 25(2), 95–113.

Wittinsglow, R. and Lehtinen, S. (2024). "The Use of Light Installations in Architectural Reconstruction: New Technological Solutions for the Public Sphere", *Routledge Companion to the Philosophy of Architectural Reconstruction*, edited by L. Giombini and Z. Somhegyi, London: Routledge.

Zangwill, N. (2023). "Aesthetic Judgment", In *The Stanford Encyclopedia of Philosophy* (Summer 2023 Edition), edited by Edward N. Zalta and Uri Nodelman, https://plato.stanford.edu/archives/sum2023/entries/aesthetic-judgment

3 Urban Aesthetics and Power

Branding Sensory Conviviality in London and Barcelona

Mónica Degen and Beatriz Guijarro Turégano

Introduction

Selling "experiences" has become a valuable currency to represent, promote, and brand cities in the 21st century. These experiences, mediated through a variety of digital and analogue media, are commissioned and disseminated by a diversity of stakeholders from public authorities, retail and service sectors, or tourism corporations (Govers and Go, 2016; Zhou and Wang, 2014). They range from the promotion of themed restaurants such as "dining in the dark" on metro adverts; activities or events in neighbourhoods advertised on Instagram or other social media to attract visitors (Boy and Uitermark, 2020); to multimodal videos such as Expedia's "Made to Travel" campaign (2022) telling us that experiences—not "stuff"—is what matters most in our lives. Experiences, often represented as convivial encounters, have become a staple feature of product branding strategies as they "represent an existing but previously inarticulate genre of economic output" (Pine and Gilmore, 1999:ix). Zukin reflects that the heightened role culture and experiences play in urban economic and spatial development lead to "[t]he look and feel of cities reflect[ing] decisions about what – and who – should be visible and what should not, on concepts of order and disorder, and on uses of aesthetic power" (1995:7). This chapter analyses the workings of aesthetic power in branding strategies of two cities: London and Barcelona, to understand the confluence of urban politics and sensations in a digital age.

In the age of intensified (digital) mediation, cities need to sell themselves on the "global catwalk" (Degen, 2008) for inward investment to tourists and residents through the sensations and experiences they promise. In a highly competitive consumer culture, the sensory cues or aesthetics that products present and elicit are unique selling points to being noticed and appeal to an evermore discerning audience. Following Böhme (2003), we use the concept of aesthetic economy to emphasise how the sensory-experiential aspects of urban life have gained prominence and are staged to shape urban imaginaries (Degen, 2008; Degen and Rose, 2022). Urban planning, design, and architecture have shifted towards the logic of the brand transforming urban spaces into brandscapes (Arantes, 2019; Klingmann, 2007) as urban administrators, city councils, politicians, and other

DOI: 10.4324/9781003335993-3

stakeholders are driven by entrepreneurial strategies to sell urban space. Such strategies range from redesigning public spaces to appeal to a particular aesthetic taste; or dramatic starchitecture commissioned to attract global media attention; to pop-up events and design interventions increasingly curated as Instagrammable moments evoking sensations that participating in these environments produce. Key to these activities are branding processes as they shape the creation of urban imaginaries, "fragments of ideas about places, be they in the form of stereotypes, images, imaginaries, slogans keywords or simple mental associations" (Vanolo, 2017:1), distributed through analogue and now, even more prominently, digital media.

In the following, we trace the workings of what we will term "urban aesthetics" through two city branding case studies. In London, we analyse the entanglement between place branding and cultural redevelopment of an area in the financial City of London: the "Culture Mile". We show how the future regeneration of its public spaces is powerfully supported by a branding strategy that emphasises sensory engagements in and digital enactments of its public life. In Barcelona, we analyse recent urban branding campaigns to examine how promoting a new city brand is based on evoking particular sensory relations partaking in its everyday public life. In both cases, we identify the representation of an idealised "sensory conviviality" as an urban imaginary. We use the term sensory conviviality to refer to the affective and sensorial side of the social in public spaces (Overing and Passes, 2000). The spontaneous "small" everyday embodied encounters with strangers and the physical urban environment support and frame our coexistence in cities: from briefly smiling at a passer-by, to sitting on a bench, smelling the falling rain while watching life go by. As Wise and Velayutham (2014:407) highlight, conviviality, living together with both human and nonhuman world "can be understood as an atmosphere and an affect", deeply linked to our sensory experiences and embodied engagements with the world around us which can be either harmonic or full of tensions. We start our analysis of the evolving and differentiated aesthetic power dynamics inherent in contemporary urban branding strategies and their effects by developing a theoretical framework to examine why focusing on urban aesthetics—sensory urban experiences—provides an important analytical lens before turning to our case studies.

Urban Aesthetics and Power

Aesthetics is a complex and multi-layered concept often discussed in academic literature from the duality of "either cognitive judgements concerned with beauty and the sublime or as embodied sensory experiences" (Hawkins and Straughan, 2015:284). However, this is an unhelpful differentiation, especially for current urban research into the aesthetic economy as we will show (see also Hawkins and Straughan, 2015). In *Sensing Cities* (Degen, 2008), I argued that cultural politics shaping the modern city had to be understood through the lens of *urban aesthetics*—referring to both the embodied sensations evoked by places and experienced differently by various social groups interacting with urban spaces and the sensory ideologies shaping them. Comparing regeneration processes across Europe in the 1990s, I identified a common trend, namely that urban regeneration is made effective through the (re)organisation of the sensory landscape and experience. We briefly summarise these arguments as they provide a theoretical framework of how urban aesthetics stratify urban power relations. We finish by discussing how the digital mediatisation of urban spaces has further intensified and altered aesthetic processes.

The central argument of *Sensing Cities* (Degen, 2008) is the "understanding of space as genuinely embodied space" (Simonsen and Koefoed, 2020:1) which calls to analyse our encounters with the world from a phenomenological inspired standpoint: we see, hear, touch, taste, or smell our way through the lifeworld. This understanding is based on the original meaning of the word *aisthesis* as proposed by Baumgarten, which refers to "the affective and felt reality of a sensible subject" (Matthews and Wall, 2023:2). We evaluate the quality of urban spaces based on the sensory-emotional perceptions they evoke in our bodies which shape how we feel, behave in, and interact with the social and physical environment (Degen and Rose, 2012). This pre-discursive feel is immediate and in situ: our sentient bodies perceive the layout of a street or square, notice textures, the overall distribution of street furniture and buildings, the uses and activities, and the general sensory feel a place offers.

We then assess and evaluate these perceptions, informed and mediated by the ways places are represented, evoked, and performed in TV, film, advertising, or social media (Rose, 2011). These representations make us familiar with places we have never physically been to before and are "central to the apprehension of cities" (Brunsdon, 2018:23). Hence, as Sennett highlights, "the city" can mean (at least) two things "one a physical place, the other a mentality compiled from perceptions, behaviours and beliefs" (Sennett, 2018: 1). The two intersect through a third dimension, namely our situated and felt embodiment where personal memories, experiences, and embodied subjectivity further strongly shape our sensory relationship and judgement of our environment. For example, a mother of a toddler looks out for steps as play and danger; or women try to walk in well-lit and lively areas when walking home late at night. Thus, each body reacts and perceives its surroundings differently depending on the activities it is involved in (Degen and Lewis, 2020; Degen and Rose, 2012) and as it is positioned within a particular "power-geometry" of place depending on social characteristics such as, for example, age, ethnicity, class, abilities, gender, or sexuality (Massey, 2012).

This leads to the second argument made in *Sensing Cities* (Degen, 2008), namely that sensory perceptions are shaped by cultural values and underpin social hierarchies and distinctions (Howes, 2005). The moral judgements that we link to sensory experiences are not subjective but reflect existing power relations in society. Indeed, each society produces its own sensory hierarchies linked to an established social order where the dominant group will be linked to the positively valued senses, and the subordinate group associated with what are regarded inferior and polluting senses (Smith, 2006). This highlights the importance of researching the "social ideologies conveyed through sensory values and practices" (Howes, 2005:4) and reveals how power is transmitted and experienced somatically through lived experience and cultural practices, representations, and urban imaginaries.

The third argument in *Sensing Cities* (2008) is that aesthetics, sensory experiences, are key elements of contemporary competitive urban development and policy used to attract, deter, and shape people's uses and attachments to place. Since the 1980s, urban regeneration projects have consciously concentrated on transforming the aesthetic features of places from functional landscapes to "an architecture of spectacle, with its sense of surface glitter and transitory participatory pleasure, of display and ephemerality, of jouissance" (Harvey, 1989:91). Consumer culture has radically transformed urban environments creating spaces of "multimodal experience, not just that of vision, a space of virtualities and intensities that actualise themselves not as objects but as events" (Lash and Lury, 2007:15). Böhme's discussion of the aesthetic economy as producing, staging, and selling sensory pleasures, or atmospheres, to support contemporary urban redevelopment resonates here as he identifies:

> a new resulting aesthetics [...] concerned with the relation between environmental qualities and human states. This "and", this in-between, by means of which environmental qualities and states are related, is atmosphere.
>
> (Böhme, 1993:114)

Indeed, the case studies analysed in *Sensing Cities* (2008) Castlefield in Manchester and el Raval in Barcelona illustrate how their regeneration strategies drew on comparable spatial-sensory techniques in the late 1990s to stage-specific atmospheres (Degen, 2008). These consisted of sandblasted heritage features mixed with minimalist urban design features that provided visual stages for events, yet lacked other sensory intensities. They were experientially interchangeable regenerated spaces, creating what Rancière (2006) would describe as a particular "sensory fabric", constructed around the sensory values or order, purity, and cleanliness (Degen, 2008).

Rancière, in his conceptualisation of aesthetics as the "distribution of the sensible" highlights how sensory perceptions frame our common-sense understandings of places and society (Blakey, 2021). His work is key to understanding how political power operates precisely through

> politics as a form of experience. Politics revolves around what is seen and what can be said about it, around who has the ability to see and the talent to speak, around the properties of spaces and the possibilities of time.
>
> (Rancière, 2006:13)

As Rancière suggests, particular sensory regimes are implemented by the state or corporate institutions that create "the configuration of a perceptive field" (Dikec, 2015:92) and thereby influence and control who or what can be perceived, heard, seen, touched, tasted, or heard. For example, in the case of el Raval and Castlefield's regeneration, middle-class aesthetic codes were integrated into the physical fabric of these working-class neighbourhoods through a multiplicity of cultural regeneration policies and strategies. These ranged from applying a "designer heritage aesthetic" to symbolically recoding these neighbourhoods into cultural quarters; the opening of museums, universities, and art galleries or staging cultural and fashion events to attract particular audiences (see Degen, 2008). However, the distribution of urban sensations is never hegemonic or complete, even if it may appear as such. For example, Summers (2021) examines contestations around sound aesthetics in a gentrifying Black neighbourhood. She shows how local, African-American go-go music is under threat of being muted and pushed out by gentrification, yet Black residents reclaim this soundscape to assert their identities in relation to place. The anti-gentrification movement #DontMuteDC reclaims space and resists gentrification and displacement by playing music on the streets. Summers shows the power of Black sound aesthetics in reclaiming areas they have been pushed out from and describes this as the successful logic of "reclamation aesthetics" to take back and to "[assert] that Black people not only have the right to stay in the city, but also should move fluidly through city spaces" (2021: 43).

Our discussion has shown so far that urban aesthetics is at the centre of urban life and politics as sensory power relations are imposed, played out, and resisted on different scales. It is helpful here to remind us of the different yet linked understandings of aesthetics we refer to in this chapter. First, aesthetics is understood from a phenomenological inspired perspective, which refers to directly experienced sensations. And, second, aesthetics in Rancière's understanding is "a shared mode of aesthetic engagement" (Ghertner, 2015:7) that becomes common-sense and is ideological as exemplified by the way urban redevelopment and policy adopts particular aesthetic regimes. For example, Ghertner (2015), similar to Degen (2008), shows how an "aesthetic governmentality" has become common sense in Delhi's aim to become a world-class city as planning strategies and processes. For example, certain sensory experiences are portrayed as desirable such as the creation of designer shopping malls and others, negative, such as Delhi's slums. This "rule by aesthetics", the" process of translating broad aesthetic codes into a governing lens for organizing urban space" (Ghertner, 2015:125), has real effects on the ground as it both transforms the city's physical environment while it simultaneously reconfigures individual's embodied sensory experiences in places.

Branding is a key "technology of perception" (Rajagopal 2001 quoted in Ghertner, 2011: 290) as it consists of constructing particular sets of associations to entice the consumption of a product, often based on an emotional connection that is established between a product and its audience (Kotler, 2002). Urban branding is distinctive in that it relies on combining both the hard, physical, and soft, sensorial, attributes of the city (Kavaratzis, 2005).

Having a distinctive urban landscape with unique natural and man-made features (from heritage buildings to enticing beaches) is as important as providing softer qualities such as engaging atmospheres in public places that sensorially envelop visitors (see Kavaratzis, 2005). This connection between hard and soft urban features is increasingly established through the selling of an urban lifestyle that emphasises what we define as sensory conviviality: being in a particular city or neighbourhood promises distinct sensory experiences and identifications with specific trends and social groups. Digital media have further intensified these aesthetic branding processes (Zhou and Wang, 2014; Banks, 2022; Degen and Rose, 2022).

Indeed, the unremitting use of digital technologies in all realms of life from smartphone use to computing has further increased the reach of aesthetic governmentality in cities affecting how urban spaces are experienced. Digital technologies have radically reframed how we live in and interact with cities, from urban transit systems that use smartcards and predict capacity and local apps to order taxis or find a restaurant to navigational devices like Google Maps. In urban environments, we are surrounded by digital screens from mobile phone screens we carry in our hands to consoles or LED screens. Kitchin and Dodge (2014) use the term "code/space" to describe spaces reliant on digital mediations. More recently, Degen and Rose (2022) have described the intimate entanglement of digital technologies and the sensorial feel of urban redevelopment as the "new urban aesthetic". From architects and designers using software to create computer-generated images to evoke particular sensations to sell future urban environments (Degen, Melhuish, and Rose, 2017); to the use of social media campaigns—digital technologies do not just reframe the urban imaginaries of cities but also its everyday embodied experiencing. In this convergence of the urban and the digital, Degen and Rose define sensory experience as central "to the reconfiguration of how cities are being branded, designed, planned, built, and lived. Digital data, software, and devices increasingly mediate how city spaces are sensed and corporeally perceived, thus reconfiguring urban experience into specific forms of urban sensibilities" (Degen and Rose, 2022:146). Yet, this new urban aesthetic is neither uniform nor hegemonic. The new urban aesthetic manifests in different ways in different urban locations: it might be glamorous or might flow or be dramatic and engages different bodies in a variety of ways (Degen and Rose, 2022).

In this chapter, we contribute to these debates by analysing the aesthetic power relations emerging from the confluence of digital mediatisation and sensory experiences in contemporary urban branding, paying particular attention to the deliberate foregrounding of sensory conviviality as an atmosphere. So far, we have outlined our theoretical framework to analyse how sensory ideologies shape and inform the branding and design of urban spaces. For the remaining, we turn to two urban branding case studies to examine the role sensory experiences play in branding cities in an ever-busier competition for digital attention.

Branding Sensory Conviviality

In this section, we analyse two examples of how branding through the senses has become a central technique to promote new urban imaginaries and shape aesthetic power relations in the city. Both case studies illustrate how an aesthetic governmentality operates through urban imaginaries that emphasise particular sensory embodied experiences while deeming others invisible - with lived effects on the ground.

The Culture Mile: Sensory Conviviality and Anticipatory Urbanism

The first case study examines the branding, visioning, and redevelopment of a new cultural district in the financial district of London. The Culture Mile was launched in 2017 in the Smithfield Market area, Britain's oldest operating meat market, as a partnership between the City of London Corporation and key cultural institutions. Its initial conception emerged from the first cultural strategy developed in 2010 by the Corporation of London, the City's municipal governing body, which aimed to "invite people to experience [the City's] cultural offerings [...] to draw in more visitors to this area" (Cultural Hub Working Party Report, 2013). Its target was to promote the City's cultural identity alongside its status as a financial centre. We now discuss the key features of this redevelopment strategy in relation to the power dynamics of urban aesthetics.

It is important to start by highlighting that the branding strategy of the Culture Mile pre-dates the redevelopment of the area—meaning that a powerful urban imaginary was constructed, mediated, and circulated independently and before any actual physical changes in the area occurred. In 2015, an urban design practice was commissioned by the Corporation of London to develop a regeneration strategy for this new cultural district, followed by the appointment in 2016 of strategic branding agencies to create a place brand which identified four key promotional/development values for the area: joined-up, generous, agile, and experimental. They developed the name "Culture Mile" and an associated suit of place branding imagery, from logos to an Instagram account, website, and promotional videos (for an extensive discussion see Degen and Ward, 2022). The logo, inspired by the brutalist aesthetic of the Barbican, was displayed throughout the neighbourhood in bright colours to visually delimit the sensory boundaries of the Culture Mile; see Figure 3.1.

In 2018, a "Look and Feel Strategy" developed by Fluid Architects "[built] upon these [branding] values to provide an approach for an unrivalled visitor experience" (2018:4) suggested a range of urban design, policy, and cultural interventions. What we are witnessing here is a merging of branding and planning, a "branding practice [that] has adopted proactive, project-driven planning rather than traditional planning approaches that were deemed passive and reactive by new municipal administrations" (Bonakdar and Audirac, 2020:150). While planning has always been future driven through visions or masterplans, a new form of anticipatory urbanism (Degen and Rose, 2022) is

Figure 3. 1 The Culture Mile (Source: Monica Degen).

emerging where branding sensory experiences in a future place earmarked for regeneration is essential to attract financial investment. As Bonakdar and Audirac criticise, this raises questions around power and social justice as place branding "remain[s] essentially a competitive urban policy for growth, free from any responsibility to the public" (Bonakdar and Audirac, 2020:150).

This "Look and Feel" (2018) strategy emphasised explicit attention on, and transformation of, the sensory physical infrastructure and sensations evoked in public spaces with four aims. First, to form a culture spine to create a "unique urban destination" with "pedestrian experience at the forefront of the public realm enhancement" and "memorable new journeys" including "green walls, striking routes and kinetic textures" (Look and Feel Strategy 2018). Second, to "[externalise] the areas world class cultural activities by embedding cultural content into the streets" (Look and Feel Strategy 2018) which includes accentuating features and textures of facades and animating public spaces with a regular programme of events. Third, "to discover and explore" by arranging a lightning strategy that visually enhances and stages some architecture, yet disregards others. Fourth, to be recognisable and to be different which emphasises the Culture Mile's physical environment as a unique selling point, emphasising the importance of social media in mediating this experience. We can see here how an aesthetic as common sense is crafted and ingrained into the urban environment to orchestrate particular sensory experiences.

This leads us to our third discussion point, namely that branding in The Culture Mile is strongly based on a public realm enhancement strategy where "Culture Mile's physical environment is its brand" (Look and Feel Strategy 2018:71). Thus, it is not the current physical spaces that are the brand but those transformed in the future, especially arrival points for prospective visitors which will be redesigned into "creative, delightful and generous" spaces to provide a sense of place and identity (Look and Feel Strategy 2018:30). The current landscape instead is described as problematic and lacking points of engagement and character—it is "out of place". We can see here how aesthetic power operates as branding is used to project a negative aesthetic, or common sense, onto an existing place experience. The new Culture Mile will surface from new forms of sensory conviviality, producing particular atmospheres stemming from a redesigned sensory landscape. As Böhme explains, the design of a building or a bench is not enough to project an atmosphere. Rather, objects need to be sensuously choreographed, arranged, and displayed in particular ways (Böhme, 2014). The transformation of the tactile and visual landscape is an important tool for recoding the meanings and identity of the existing neighbourhood as the "Culture Mile" for an external audience:

> Unity of materials, visual coherence of furniture and landscaping will play a major role in providing navigational clues that speak to users on an instinctive level. Distinctive paving can give pedestrians confidence to know they are welcome and have arrived. Public art can draw people towards key spaces, furniture can bring playfulness and an appreciation of community in public spaces, lighting can pick out landmarks and indicate where it is appropriate to walk at night and a unified planting strategy can bring to life spaces and enhance visitor memory.
>
> (Look and Feel Strategy 2018: 23)

Public space and the atmosphere of sensory conviviality it evokes, in other words its sensory experiences, new encounters fostered by a redesigned environment, and renewed public life, become the stage to express the Culture Mile's brand identity.

Yet, the planned transformations on the ground are not enough to attract new audiences. Simultaneously, a digital branding strategy needs to be developed that "stretches beyond the physical public realm" and "will reach into digital spheres such as websites, which often operate as the first point of contact for many visitors" and that enables professionals, designers, and communities "to use Culture Mile's public spaces as platforms or stages" (Look and Feel Strategy 2018:23–27). As Barns (2020:163) highlights one way social media is transforming urban branding and design is that current public art curation is moving away from installing permanent artworks and instead programming art events where the public takes active part in the performance of an urban spectacle. For example, in 2020 an art walking trail, Around the Corner, was developed for the Culture Mile to improve wayfinding, create new pedestrian experiences, and provide Instagrammable moments. The public could follow and interact with sculptures that read: "What are you going to meet if you turn this corner?" as they were guided to key neighbourhood landmarks. We can see here how physical interventions such as short-term pop-up events or interactive art interventions are not only staged to attract an online community that will visit, support, and mediate the area's events but, simultaneously, convey an anticipatory atmosphere mediated online that shows what kind of sensory conviviality needs to be performed in situ.

The Culture Mile example shows how branding, sensory recoding, and urban planning coalesce in the development of a new cultural quarter. A particular sensory conviviality is ingrained in the place through activities and branded by urban imaginaries informed by a limited set of cultural codes such as artistic, playful, and experimental and promoted digitally. The aesthetic governmentality is anticipatory, preparing the area for future development and shaping through cultural events and activities sensory experiences and encounters that create particular atmospheres aimed at specific publics: the trendy, young, affluent as well as families. This spatio-aesthetic logic frames a particular "distribution of the sensible" and shows how different institutions and organisations establish through sensory configurations what is perceived as "common-sense" (Rancière, 2006; Ghertner, 2015). The outcome is that specific sensory regimes are produced and highlighted through the branding strategy and its implementation while other sensations become invisible or less perceptible such as, for example, the area's strong working-class life and heritage, LGBTQ+ history, and club culture (for an extensive discussion see Degen and Lewis, 2020, see also Degen, 2008). This particular distribution of the sensible frames is "what is seen [perceived] as in and out of place" (Blakey, 2021:635) with tangible consequences. Indeed, the area has been designated in 2023 as a Business Improvement District with the meat market moving in 2026 to Dagenham, its buildings giving room for the relocation of the Museum of London (Degen and Ward, 2022).

Barcelona: Sensory Conviviality as a Tool to Reposition the City

Our second case study examines two urban branding campaigns of Barcelona's Tourism Board on social media channels: Barcelona Inspira (2014) and Always Barcelona (2019) which we analyse in relation to their promotional videos[1]. While much has been written since the Olympics in 1992 on "la marca Barcelona" [the brand Barcelona] (Balibrea, 2017; Delgado, 2007), Barcelona's aim to promote itself internationally is not new. Indeed, we can find traces of strategies that combine urban renewal and place promotion since the 19th century when the local bourgeoisie and industrialists aimed to promote the city as "a European city [...] with the generic objective of modernising the urban space and as an expression of an idealized vision of the city" (Monclús, 2000,

3–4) with the Universal World Exhibition of 1888. The event led to monumentalising and beautifying the public spaces of the city through public sculpture, triumphal arches, urban landscaping, various elements of urban decoration, and new pavements "to express the values and culture of the urban elite, to educate the citizens as well as to impress the visitors" (Espuche et al., 1991).

While at the start of the 20th century Barcelona carried on promoting itself to an international audience, for example as "the pearl of the Mediterranean", this came to a halt with the Civil War and Franco's dictatorship. Punished for being a city of republican and anarchist movements during the Civil War, Barcelona became for most of the 20th century a forgotten city, with little advertising exposure to the outside world, also described as its "grey period" (see Degen, 2004:132). This changed with Barcelona's nomination in 1986 by the International Olympic Committee as the host city for the 1992 Olympic Games which had never been hosted in Spain. This "flagship event" was spearheaded by the local authorities who saw it as an opportunity for "the renewal of the image of the city and the use of the events as catalysts for pursuing specific operations of urban redevelopment" (Monclús, 2003). During the 1980s, the city already used a range of local communication and advertising strategies to gain public support for urban renewal and modernisation of public spaces. Some of the campaigns such as "Barcelona mes que mai" [Barcelona more than ever] (1984), "Barcelona posat guapa" [Barcelona get pretty] (1985), or "Visca Barcelona" [Long live Barcelona] (2008) were all aimed at the city's citizens to promote a sense of pride similar to other campaigns in cities across Europe with the internal aim linked to the reconstruction of the city as an ideal place to live (Degen, 2004; Balibrea, 2017). While there were international tourism campaigns during the Olympics, there was no specific urban branding strategy. It was really only after the Olympic Games that the international projection the city had achieved became clear.

At the start of the 21st century, the policy to brand locally changed to a more strategic global focus. A 2011 study by Barcelona City Council revealed that while the city had become a tourism hotspot, it was "still not globally recognized for its businesses, services, innovation, knowledge and entrepreneurship"; thus, a new brand strategy had to be conceived (Belloso, 2015). One of the first communication campaigns for outward tourism and investment was "Barcelona Inspires" (2014) which aimed to reposition Barcelona away from being a tourism venue to being a place for entrepreneurial talent, international business and residence as well as pro-active economic activity. The video campaign features a montage of everyday moments to stage, in Böhme's sense, particular atmospheres of the city, with its citizens as its key protagonists, rather than well-known tourist venues or famous buildings. Instead, we see a collage of Barcelona's nightscape with illuminated buildings and moving traffic, people or clouds interspersed with close ups of moments of conviviality such as a mother playing in a playground with her daughter on a seesaw; a Barca fan shouting at a football game in Camp Nou, or a shopkeeper pouring coffee beans into a bag. Each frame contains annotated graphics that suggests the sensations evoked such as, for example, the warm air in the playground, the shouting of the football fan, or the smell of coffee beans. In the background, one can hear repetitive upbeat music suggesting modernity and activity while a male voice invites the viewer to immerse themselves into the sensory experiences featured:

> Stop for a second. Take time to open the window. With the sky, the street, the other windows, here is something special in the air. Something that you quite don't know what it is. Some say it's the sun, others the sea...signature cuisine, design, architecture...it may also be the people, their creativity or their good sense...Maybe because it is a modern, cosmopolitan city... Barcelona has all this, but there's something else that you feel everyday. But that no-one...knows what it is for sure. Barcelona, a capital that inspires!

We can see in this example how aesthetic governmentality works through a branding campaign that evokes embodied engagements in and sensory stimuli emanating from specific activities, places, and features—those supporting innovation, creativity, and entrepreneurship and featuring mainly an affluent white public. The video represents an idealised urban imaginary, constructing a common-sense aesthetic, reflecting decisions about what—and who—can be sensed or not (Zukin, 1995:7). Barcelona's ethnic diversity for example is only hinted at by a group of Black performers in a theatre; the places featured are all in the centre of the city—unaffordable for most of the population. A specific sensory conviviality is represented that emerges from being in affluent and exclusive spaces such as a designer restaurant, an expensive flat, or a theatre that stages a "Barcelona Inspires" atmosphere that most residents of the city would not be able to access.

In October 2014, Barcelona's city council created the agency: "Barcelona Brand" as a private–public enterprise that oversees and coordinates the diverse actions and practices of the city's urban branding. In 2019, just before the pandemic, it launched a new phase of the Barcelona brand. The council was concerned by the repercussions that two major crises had on the city's international brand: the terrorist attacks on Barcelona's main boulevard Las Ramblas in August 2017 and the protests and police brutal repression of 1 October 2017 that saw the city in flames across global media. In 2018, a "Brand Book for Barcelona" was produced by local authorities "to construct a global, integrated perception of the city" (Ayuntament the Barcelona 2018:7), based on a perception study of the city. The campaign "Always Barcelona" (2018/19) emerges from these descriptions to "consolidate its international reputation and project[...] the image of a vital and strong city, a magnet for the generation of economic activity" (Ayuntament the Barcelona 2018:7). Similar to Ghertner's (2015) discussion of aesthetic governmentality based on codes of appearance in Delhi to produce a world-class city, local authorities are engaging in careful perception management and crafted imagining to organise a "community of sense" (Rancière, 2006), "a shared mode of aesthetic engagement" (Ghertner, 2015:7) to define how the city should be emotionally and experientially evoked through branding.

The campaign that emerged consisted of a digital branding campaign designed by the Catalan company Folch that showed a different perspective of the city, one that narrated "Barcelona through the lives of its citizens" (Folch, 2018) and portrayed everyday sensations generated by living in the city:

> We wanted to represent the city as a place that is alive with opportunities for people coming to live, work, or even set up a company. When you think about a city you think about the people you know that live there, the music or culture that you experienced, or the little cafe on the corner that you love. It's a very personal thing – a feeling based on fragmentary individual moments. This is what #AlwaysBarcelona tries to capture.
>
> (Folch, 2018: np)

The campaign had no logo but was instead conceived as a social media campaign in three stages building on each other and based on international events: first "Share Like Follow" focuses on the everyday life in Barcelona and is targeted to the Mobile Congress; second "Think Design Create" focuses the creative talent and design industry, launched for Barcelona's Design week; and lastly, "Play Act Dance" targeting the Sonar music festival audiences and featuring the city's underground music scene. The campaign had a strong digital emphasis concentrating on both organic and paid social media, using geotargeting to reach audiences and focusing on digital press-sites, Facebook, and Instagram carousels as well as YouTube videos (Folch, 2018).

For the purpose of this chapter, we focus on the "Share Like Follow" video campaign[2] that uses a well-known digital communication slogan by YouTube or Instagram influencers. The video has no voiceover; instead, it starts with upbeat urban electronic music by the local band Barcelona Lasers and depicts a range of embodied encounters. The video begins with a port worker taking a picture on his mobile of the sunrise in a harbour, followed by a montage of moments of sociability in Barcelona's public spaces such as a young mixed-race bohemian female shop owner cleaning her window and smiling; group of ethnically diverse youth trying out a choreography in front of museum windows; older Catalan men playing passionately domino on an improvised table on the beach; and two young white women kissing on the Ramblas. The video finishes with two young people jumping into the Mediterranean Sea at sunset and the slogan "Share Like Follow" appearing in white superimposed on the image, finishing with the words Barcelona across the screen. Most locations in the video are not identifiable to outsiders, it consciously does not show well-known tourist locations. Instead, it is moments of sensory conviviality by ordinary citizens and visitors dancing together, working, kissing, and exercising that become the centre stage. We encounter the city through the lens of an idealised sensory conviviality. The lived experiences of ordinary citizens operate as signifiers for

Barcelona as an open, liberal, and multiracial urban environment as suggested in its brand book, as a place to work, live, experience, and most importantly, to invest in. As Böhme suggests:

> things and their form is not what is at stake. Rather, the focus on scenes, life spaces, charisma. Here, atmosphere is the explicit object and goal of aesthetic action. […] what matters is their radiance, their impressions, the suggestions of motion.
>
> (2014:45)

Thus, the staged scenes are carefully assembled, choreographed, and filmed to produce and sell an atmosphere: ordinary moments of sensory conviviality are staged as a lifestyle for the global consumer classes.

The use of a collage of snapshots, the format, the colours, and the overall sensory vibe is deliberate and influenced by the power of social media platforms as the creators themselves acknowledge:

> We needed to adapt our voices to the format and the audience we were talking to. To create empathy and make sure our message gets through. So, what we did was apply the mobile-first commandment, writing the copy directly in the iPhone Notes app so it would be built in a native environment, emojis and all, in the same way in which the assets would be spread.
>
> (Folch, 2018)

We can see here how digital platforms such as Instagram not only support the global sharing of images but shape the themes, narratives, and aesthetic look of branding campaigns such as choosing everyday moments of conviviality to create particular dramatic images (Degen and Rose, 2022).

What do these branding strategies tell us about the workings of aesthetic power? Summers (2019) argues in regard to the branding of Black aesthetics that "blackness" becomes an aesthetic infrastructure and trope of cool enveloped in a discourse of diversity. This leads to a conscious avoidance of discussing social inequalities or commitment to changing racist policies and actions. Similarly, we witness in this branding of Barcelona very selective moments and tropes of a particular form of cool-bohemian, multiracial, and middle-class sensory conviviality that portray a city without conflict, poverty, or problems for the sake of selling desirable urban products as part of wider neoliberal processes in the city. This has had a noticeable impact shaping the everyday life of Barcelona's citizens from rising rents, to protests around the touristic gentrification of neighbourhoods with AirBnB flats, or protests by local residents against overtourism (Mansilla and Milano, 2019). The negative repercussions for the local populations' sensory conviviality are intensifying as Cocola-Gant (Cocola-Gant and Lopez-Gay, 2020; Cocola-Gant, 2023) argues, not only from the overcrowding of public spaces by tourists or the replacement of local shops and

Figure 3.2 #Always Barcelona (Source: Arxiu Folch photography by Leo García Méndez).

restaurants for tourist ones. More interestingly, Barcelona has witnessed in recent years the rise of "foreign only" enclaves in centrally located neighbourhoods fostered by international lifestyle migrants. As Cocola-Gant (2023) concludes such transformations are leading to an ongoing feeling of loss, expulsion, and a deterioration of quality of life, even mental distress for local residents.

Conclusion

By examining urban branding strategies in London and Barcelona, this chapter has illustrated how aesthetic power dynamics underpin the construction of urban imaginaries. We have shown how branding operates as part of a broader aesthetic governmentality in which authorities aim to shape a common-sense aesthetic by reconfiguring urban imaginaries of places. In both case studies, embodied sensory experiences mediate an idealised sensory conviviality that becomes an aesthetic tool through which urban development or the repositioning of an urban brand gets instigated. We have demonstrated that these imagined experiential constellations are selective and engineered as tools to brand, mediate, and ultimately shape urban environments for particular social groups, practices, and lifestyles while leaving others out. Aesthetics is an important device shaping social in- and exclusions in the contemporary city.

Both case studies are public authority branding strategies employing external consultants and architects. Utilising commercial agencies by their nature tends to favour a more commercial emphasis on urban life, where concerns around welfare and equity become ever less imperative. In the example of the Culture Mile, we analysed how branding, the framing of sensations, and urban redevelopment go hand in hand. The area gets branded before actual redevelopment takes place. This anticipatory urbanism works by recoding the areas' public spaces sensorially through specific brand identity values: artistic, generous, playful, and experimental—which explicitly inform pop-up art and events that stage particular sensory convivialities aimed for consumption by an affluent, trendy, and bohemian public. Aesthetic governmentality is implemented through public space animation decisions that choreograph what kind of sensory experiences are worth staging and make particular atmospheres tangible while concealing others (Ghertner, 2015). In our second example, we examined Barcelona's recent branding strategies to reposition the city as an entrepreneurial place to work and invest in rather than a tourist venue. We argue that in the urban imaginary constructed, it is the city's sensory conviviality—the embodied sensory engagements with others in public spaces—that is the main protagonist of these campaigns. A particular configuration of sensory conviviality is presented through limited ways of experiencing, practices, and lifestyles that represent a modern, liberal, and cosmopolitan city while many other experiences are left out. Both case studies further illustrate how the development of digital technologies, in particular social media, increases the importance of sensory experiences to successfully mediate these branding campaigns, determining not only the dissemination but also the format, content, and focus of urban branding strategies. For example, in the Culture Mile physical spaces are designed to become Instagrammable moments, while in Barcelona the branding videos are filmed based on a social media aesthetic to evoke a "rich, authentic picture" (Folch, 2018).

The aim of this chapter has been to illustrate why aesthetics, as an analytical tool, is key to making sense of the city in an aesthetic economy (see also Degen and Rose, 2022; Matthews and Wall, 2023; Summers, 2021). To this end, we have suggested a broader, more encompassing, notion of aesthetics that brings together Rancière's understanding of aesthetics as "common-sense", where "we are bound together by a 'sensory fabric' that shapes how we make sense of the world" (Blakey, 2021:624) with an understanding of aesthetics as embodied sensory experiences. Doing so links the embodied, felt nature of representations with ideological values. It helps to reveal how sensory power works through the staging and framing of atmospheres. An atmosphere is generated as a relation of multiple elements and personal memories, memories of films, social media posts, and branding campaigns that further punctuate urban spaces (Degen and Rose, 2022). As we have shown, sensory conviviality is at the centre of both branding campaigns as *the* desirable atmosphere.

Rancière (2006) argues a particular social order is maintained through an established set of possible modes of perception. While specific aesthetic regimes of experience are constructed and established as common sense through the urban imaginaries represented by branding campaigns, alternative or new modes of sense perception to make visible different people, stories, and knowledges are excluded. Who and what is sensed and whose sensations are represented and valued socially and economically stratify society (Rancière, 2006). As we indicated the branding strategies are part of wider neoliberal planning processes transforming the urban environment, the Culture Mile has been designated a Business Improvement District; in Barcelona, some areas are attracting more foreign investors and residents than local citizens. As Lindner and Sandoval remind us "aesthetics are increasingly being used [...] to produce seductive conditions and instil desires needed for creating exclusionary urban transformations predicated on displacing and disempowering vulnerable populations" (Lindner and Sandoval, 2021:15). However, as Summer's discussion of resistance by Black people against displacing and erasing go-go music from the cultural and sensory geography in Washington illustrates, the power of urban aesthetics is never absolute but always open for contestation: "activism [in itself] is an aesthetic process essential to contestations over the organization of society, as actors mobilise radical forms and spaces of expression" (Summers, 2021: 33).

Notes

1 For Barcelona Inspira see: https://www.youtube.com/watch?v=H9ZqclVIsxk.
2 https://www.youtube.com/watch?v=MPkE0C6BwB0.

References

Arantes, P. (2019). *The Rent of Form: Architecture and Labour in the Digital Age*. University of Minnesota Press.
Ayuntament de Barcelona (2018). Identitat i posicionament de Barcelona. Un relat coral. Barcelona: Ajuntament de Barcelona. Available at: https://bcnroc.ajuntament.barcelona.cat/jspui/handle/11703/112549 Accessed 17 Apr. 2025.
Balibrea, M. P. (2017). *The Global Cultural Capital*. London: Palgrave MacMillan.
Banks, D. A. (2022). The attention economy of authentic cities: How cities behave like influencers. *European planning studies*, 30(1), pp. 195–209.
Barns, S. (2020). *Platform Urbanism: Negotiating Platform Ecosystems in Connected Cities*. Basingstoke: Palgrave Macmillan.
Belloso, J. C. (2015). Rebranding Barcelona as a City for Business, Talent and Innovation: City Branding Case Study, The Place Brand Observer, https://placebrandobserver.com/rebranding-barcelona-city-branding-case-study/ accessed 24/2/2023.
Blakey, J. (2021). The politics of scale through Rancière. *Progress in Human Geography*, 45(4), pp. 623–640.
Böhme, G. (1993). Atmosphere as the fundamental concept of a new aesthetics. *Thesis Eleven*, 36(1), pp. 113–126.
Böhme, G. (2003). Contribution to the critique of the aesthetic economy. *Thesis Eleven*, 73(1), pp. 71–82.
Böhme, G. (2014). Urban Atmospheres: Charting New Directions for Architecture and Urban Planning. In *Architectural Atmospheres: On the Experience and Politics of Architecture*, edited by C. Borch. Basel: Walter de Gruyter GmbH, 42–59.

Bonakdar, A and Audirac, I (2020). City branding and the link to urban planning: Theories, practices, and challenges. *Journal of Planning Literature*, 35(2), pp. 147–160.

Boy, J. D. and Uitermark, J. (2020). Lifestyle enclaves in the Instagram city? *Social Media + Society*, 6(3), 1–10.

Brunsdon, C. (2018). *Television Cities: Paris, London, Baltimore*. Durham: Duke University Press.

Cocola-Gant, A. (2023). Place-based displacement: Touristification and neighborhood change. *Geoforum*, 138, pp. 1–10.

Cocola-Gant, A. and Lopez-Gay, A. (2020). Transnational gentrification, tourism and the formation of 'foreign only' enclaves in Barcelona. *Urban Studies*, 57(15), pp. 3025–3043.

Cultural Hub Working Party (2013). *Cultural Hub Working Party Report to Policy and Resources Committee*. London: City of London Corporation.

Degen, M. (2004). Barcelona's Games: the Olympics, urban design, and global tourism. In *Tourism Mobilities*, edited by Mimi Sheller and John Urry. London: Routledge, pp. 143–154.

Degen, M. (2008). *Sensing Cities: Regenerating Public Life in Barcelona and Manchester*. London: Routledge.

Degen, M. and Lewis, C. (2020). The changing feel of place: the temporal modalities of atmospheres in Smithfield Market, London. *Cultural Geographies*, 27(4), pp. 509–526.

Degen, M., Melhuish, C. and Rose, G. (2017). Producing place atmospheres digitally: Architecture, digital visualisation practices and the experience economy. *Journal of Consumer Culture*, 17(1), pp. 3–24.

Degen, M. and Rose, G. (2012). The sensory experiencing of urban design: The role of walking and perceptual memory. *Urban Studies*, 49(15), pp. 3271–3287.

Degen, M. and Rose, G. (2022). *The New Urban Aesthetic: Digital Experiences of Urban Change*. London: Bloomsbury Publishing.

Degen, M. and Ward, I. (2022). Future Urban Imaginaries: Placemaking and Digital Visualizations. In *Seeing the City Digitally*, edited by G. Rose. Amsterdam: Amsterdam University Press.

Delgado, M. (2007). *La ciudad mentirosa*. Barcelona: La Catarata.

Dikec, M. (2015). *Space, Politics and Aesthetics*. Edinburgh: Edinburgh University Press.

Espuche, A. G., Guardia, M., Monclús, F. J. and Oyón, J. L. (1991). Modernization and urban beautification: The 1888 Barcelona World's Fair. *Planning Perspective*, 6(2), pp. 139–159.

Folch (2018). The Reputation Campaign #ShareLikeFollowBCN by Folch [online]. Available from: https://www2.folchstudio.com/sharelikefollowbcn/ (Accessed 23/03/2023).

Ghertner, A. (2015). *Rule by Aesthetics: World-Class City Making in Delhi*. New York: Oxford University Press.

Ghertner, D.A. (2011). Rule by Aesthetics: World-class City Making in Delhi. In *Worlding Cities: Asian Experiments and the Art of Being Global*, pp. 279–306.

Govers, R. and Go, F. (2016). *Place Branding: Glocal, Virtual and Physical Identities, Constructed, Imagined and Experienced*. London: Springer.

Harvey, D. (1989). *The Condition of Postmodernity*. Cambridge, MA & Oxford: Blackwell.

Hawkins, H. and Straughan, E. (2015). *Geographical Aesthetics: Imagining Space, Staging Encounters*. London: Routledge.

Howes, D. (2005). *Empire of the Senses*. Oxford: Berg Publishers.

Kavaratzis, M. (2005). Place branding: A review of trends and conceptual models. *The Marketing Review*, 5(4), pp. 329–342.

Kitchin, R. and Dodge, M. (2014). *Code/space: Software and Everyday Life*. MIT Press.

Klingmann, A. (2007). *Brandscapes: Architecture in the Experience Economy*. MIT Press.
Kotler, P. (2002). *Marketing Places*. New York: Simon and Schuster.
Lash, S. and Lury, C. (2007) *Global Culture Industry*. Cambridge: Polity Press.
Lindner, C. and Sandoval, G. F. (eds.) (2021). *Aesthetics of Gentrification: Seductive Spaces and Exclusive Communities in the Neoliberal City*. Amsterdam: Amsterdam University Press.
Look and Feel Strategy (2018). *Culture Mile: Look and Feel Strategy*. London: City of London Corporation. Available at: https://www.cityoflondon.gov.uk/assets/Services-Environment/public-realm-culture-mile-look-and-feel-strategy.pdf [Accessed 17 Apr. 2025].
Mansilla, J. A. and Milano, C. (2019). Becoming centre: Tourism placemaking and space production in two neighbourhoods in Barcelona. *Tourism Geographies*, 24(4–5), pp. 599–620.
Massey, D. (2012). Power-geometry and a Progressive Sense of Place. In *Mapping the Futures*. London: Routledge, pp. 75–85.
Matthews, D. and Wall, I. (2023). Legal aesthesis: Affect, space and encounter. *Law, Culture and the Humanities*, 19(2), pp. 186–190.
Monclús, F. J. (2000). Barcelona's planning strategies: From 'Paris of the South' to the 'Capital of West Mediterranean'. *GeoJournal*, 51, pp. 57–63.
Monclús, F. J. (2003). The Barcelona model: And an original formula? From 'reconstruction' to strategic urban projects (1979–2004). *Planning Perspectives*, 18(4), pp. 399–421.
Overing, J. and Passes, A. (2000). *The Anthropology of Love and Anger: The Aesthetics of Conviviality in Native Amazonia*. London: Routledge.
Pine, B. J. and Gilmore, J. H. (1999). *The Experience Economy: Work Is Theatre and Every Business a Stage*. Boston: Harvard Business School.
Rancière, J. (2006). *The Politics of Aesthetics: The Distribution of the Sensible* (trans. G. Rockhill). London: Continuum.
Rose, G. (2011). *Photography and its Circulations*. London: Routledge.
Sennett, R. (2018). *Building and Dwelling: Ethics for the City*. Farrar, Straus and Giroux.
Simonsen, K. and Koefoed, L. (2020). *Geographies of Embodiment*. London: Sage.
Smith, M. M. (2006). *How Race Is Made: Slavery, Segregation, and the Senses*. North Carolina: Univ of North Carolina Press.
Summers, B. T. (2019). *Black in Place: The Spatial Aesthetics of Race in a Post- chocolate City*. North Carolina: UNC Press Books.
Summers, B. T. (2021). Reclaiming the chocolate city: Soundscapes of gentrification and resistance in Washington, DC. *Environment and Planning D: Society and Space*, 39(1), 30–46.
Vanolo, A. (2017). *City Branding: The Ghostly Politics of Representation in Globalising Cities*. London: Routledge.
Wise, A. and Velayutham, S. (2014). Conviviality in everyday multiculturalism: Some brief comparisons between Singapore and Sidney, *European Journal of Cultural Studies*, 17(4), pp. 406–430.
Zhou, L. and Wang, T. (2014). Social media: A new vehicle for city marketing in China. *Cities*, 37, pp. 27–32.
Zukin, S. (1995). *The Culture of Cities*. Cambridge, MA: Blackwell.

4 Aesthetics as Sensory Experience and the Embodied Practice of Bicycle Delivery Work

Josh Widera

Introduction

"Our worst day on the bike is better than their best day in the office", Andy winks at me, catching his breath. Sweat is glistening on our arms and faces, despite the cold night air. I just trailed the seasoned Scottish bicycle courier to a fourth-place finish at an "alleycat" at the European Cycle Messenger Championships. Trying to recover from the final sprint after two hours of racing through the streets of Edinburgh, I collapse onto my handlebars. Andy is standing upright, stretching his back, smiling in exhaustion when one of the organisers, Eva, recognises him and walks over for a chat.

Andy had been a bicycle messenger for most of his adult life and it was easy to see that he loved the profession as well as the culture and community that comes with it. This was my first introduction to the world of messengers, the "messlife", as it is referred to by insiders, and also the first time I heard the mantra that many riders have since repeated to me. The phrase is a common slogan within the subculture and encapsulates some of the pride and joy many messengers take from their work. *Our worst day on the bike is better than their best day in the office*. It conjures heroic adventure, legitimising the precariousness and hardships of the job. And it implies a common identity for bicycle couriers against those outside the community.

This chapter presents aesthetics as a minor theory (Katz, 1996) to complement an explanation of how meaning and knowledge is produced for bicycle messengers in their environment. It does so by building on a critical tradition of aesthetic theory which returns to the original meaning of aesthetics as the sensory perception of the world. In turn, this tradition offers "anaesthetics" as the depletion of embodied ways of knowing through processes of desensitisation. While some research highlights the potential benefits anaesthetics offers as a protective reaction that may be employed as a tactic of resilience (Bissell, 2022), this chapter focuses on aesthetics as the potential for embodied knowledge and experience which offers an explanation for a subject's sense of place and sense of self. Contributing to cultural geographic literature on affect and corporeal experience, this chapter illustrates how bicycle messengers produce

DOI: 10.4324/9781003335993-4

an aesthetic—a sensory experience of their lifeworld—through the embodied practice of delivery work and in spite of its precarity, alienation, and danger.

This chapter will provide a brief history of bike messengers, which sketches the different variations of bicycle-powered delivery work. The second part will introduce critical theory on aesthetics and anaesthetics and their contemporary successors in geography. Understanding aesthetics as linked to an embodied knowledge of the world, and anaesthetics as a process of corporeal desensitisation, this tradition provides a framework that helps situate why many bicycle messengers positively embrace their work as enjoyable and meaningful. Two vignettes taken from interviews and ethnographic research with bike couriers will serve to make this point in the third section. The examples from Switzerland and the United Kingdom show two messengers who profoundly enjoy and identify their work and how this is connected to kinaesthetic enjoyment and an appreciation for their own practice as craftwork. The chapter will end with a short remark on minor theory as an analytic lens, as well as a conclusion that reveals some of the negative consequences of couriers' positive embrace of their own labour.

A Short History of Bicycle Delivery Work

Bicycle delivery workers are no new phenomenon. In fact, they have existed as long as the technology of the bike itself. In the United States, at the turn of the century, messengers were a symptom and result of industrial urbanism and connected the nationwide telegraph offices with their clients: messages were brought to offices to be wired to locations around the country and telegrams were picked up and delivered to their recipients (Downey, 2002). While these workers were bound to the fate of the telegraph and faded with the rise of telephones in the 1940s, other forms of bicycle-powered delivery work continued to arise. In 1962, for example, bicyclists employed by Western Union played a part in the Cuban Missile Crisis delivering messages from the USSR's embassy to the White House (Dizard, 2001: 101). In the 1970s, bicycle couriers worked for the film industry to deliver rolls between sets and studios. And since the 1980s, with the rising importance of global finance, bicycle messengers have formed a part of the world's post-industrial cities. One of the effects of globally mobile capital has been renewed urban concentration and the revitalisation of city cores in places that also became centres of global messenger culture (Kidder, 2017: 10). Bike couriers are an integral part of those cities' communication infrastructures, connecting businesses and financial institutions, advertisement companies and their clients, architects and city planners, and attorneys and courts. Traffic congestion has continued to help messengers be a quick and convenient option to deliver in city centres (Fincham, 2007: 191).

The past ten years have seen yet another evolution of bicycle delivery. As digitisation has chipped away from the need to physically transport messages and documents, the demands for the delivery of other kinds of cargo have increased. Many couriers today have adapted to these shifts: sling bags with a single shoulder strap are replaced by bulky rolltop backpacks or the insulated cubes emblematic of food delivery apps. Fixed-gear track bikes are swapped for cargo bikes or electric bicycles to respond to changes in items' size and weight. Especially notable is the arrival and growth of platform-mediated delivery work. Since the 2008 financial crisis, labour conditions are increasingly shaped by what has come to be defined as platform urbanism (Barns, 2019; Caprotti et al., 2022), highlighting the ways in which urban capitalism relies on platforms (Srnicek, 2017) for immediate spatial and technological fixes to infrastructural and economic problems (Sadowski, 2020: 450). Across each of these historically distinct iterations, however, bicycle delivery workers are practising a more or less constant activity: the time-sensitive delivery of goods and messages.

This activity was celebrated by Andy and other messengers that evening in an "alleycat", a race across multiple checkpoints throughout the city, simulating the work of a courier. These races are indicative of a level of identification and pride that messengers feel for their job that is disproportionate to the conditions of the work. Even their spare time is often spent with friends and colleagues participating in the thriving subculture that surrounds the scene: get-togethers after work, concerts, parties, weekend rides, bike polo, and various kinds of messenger races. Kidder calls those riders who tailor their leisure activities around their job "lifestyle messengers" (2017: 52–55). They are actively participating in and producing an entire cultural sphere surrounding bicycle couriers that globally connects communities beyond the cities' limits where their labour takes place. The couriers' solidarity network is so strong that when messengers visit another large city, they can almost be certain that someone will offer them a place to stay, a bike, and perhaps even a few shifts at a local company.

This cultural field has yielded an abundance of media and goods over the years, from race footage and YouTube videos, to documentaries and magazines, to zines, novels, and non-fiction, (Day, 2015; Mendelsund, 2021; Widera, 2021), to bags and apparel. Even Hollywood has eternalised the bicycle messenger on the big screen in movies such as *Premium Rush* in 2012 and *Quicksilver* in 1986. But to many within the scene, nothing is a bigger expression of courier subculture than the national, continental, and world championships.

The first Cycle Messenger World Championships (CMWC) were held in 1993 in Berlin and have since—with a few exceptions—occurred annually. Since then, the CMWC have taken place in cities central to global messenger culture such as London in 1994, San Francisco in 1996, New York City in 2005, and Tokyo in 2009, but also smaller communities, including Edmonton in 2004, Panajachel in 2010, and Jakarta in 2019. The event, which attracts around 500 participants each year, is an opportunity to build friendships and solidarities with workers from far-flung countries and an important cultural experience for many lifestyle messengers. In a festival-like atmosphere, couriers compete in a variety of races and skill challenges, many of which test the abilities required for working as a courier, highlighting the connections between job and culture.

This thriving cultural space and global network stand in stark contrast with the conditions of the job. While these differ not only from city to city but also from company to company, couriers are usually poorly paid and rarely enjoy significant rights or protections. More often than not, they are classed as independent contractors, finding themselves in quasi-self-employed situations that allow the companies to pay their riders per delivery. This also means that messengers often lack health insurance, holiday pay, and minimum wage, and have to provide bikes and other equipment. Since their pay is usually directly dependent on how many deliveries they are able to complete, couriers take higher physical risks. This has gone so far as for some researchers to suggest that the death of riders is folded into the economic calculations of the benefitting companies (Orr et al., 2023). Sometimes, messengers report, they are assigned deliveries that are impossible to complete in time unless they "defy either: the rules of traffic or the rules of physics" (Widera, 2021: 64). While many couriers' employment situations today meet common markers of the gig economy (Woodcock and Graham, 2020) and platform capitalism, many dimensions such as independent contracts and payment per delivery predate these recent frameworks. What can be observed, rather, is in line with "deregulation and the pursuit of more vulnerable workforces", which is fundamental to neoliberal dynamics from the 1980s onwards (Harvey, 2010: 159).

In simple terms, courier work might appear to be a straightforward example of surplus value extraction. A messenger picks up an item at one location and delivers it to another but is paid less than what the client pays for the delivery. The produced surplus value is held by the courier company or delivery platform. It is not hard to see where profits are extracted by the logistics company in this model. The labour conditions for delivery work follow the same logic as so many other modes of employment in their movement towards what Karl Marx called the self-valorisation [*Selbstverwertung*] of capital: the greatest possible production of surplus value through the greatest possible exploitation of labour power (Marx and Engels, 2008: 350). The production and distribution of a commodity or message is highly compartmentalised and the courier's task in this process only extends to a few kilometres through the city. The overall deskilling involved further diminishes the value of labour power (Harvey, 2010: 181).

It would not be surprising if the work would be perceived as highly alienating, creating a sense of removal from the product of one's labour. The communicational processes of the city's information infrastructures, of which couriers form a part, are external to them; and more often than not they do not know what they are delivering. Despite the fact that most messengers would actually argue that their navigation of urban traffic, their routing, and their bike handling all require significant nuance and skill, they are a workforce under constant threat of replacement. Bicycle messengers know all this. They do not need a sermon on value extraction to understand the exploitation they experience. Yet their pride, subculture, and identification seem viscerally at odds with the sense of alienation that might be expected. I argue that aesthetics offers a productive way to better understand how bicycle messengers produce meaning and identity in these conditions. As their work contains an element of embodied practice, couriers are able to produce an affective sense of enjoyment and purpose in spite of exploitation and dangerous working conditions. Making the case for such an aesthetic understanding contributes to recent geographical interest in aesthetics and affects as important dimensions of spatial experience.

Aesthetics as Embodied Experience

To see exactly how aesthetics comes to bear on questions of alienation and embodiment, it is helpful to remember the term's origins in the Greek *aisthesis* [*αἴσθησις*]. Aesthetics was first "a discourse of the body" (Eagleton, 1990: 13) referring to the embodied experience of perception. Originally, it related more to the field of instincts and worldly sensations than to art, culture, the imaginary, and abstract judgment. Terry Eagleton shows how throughout its modern iterations, aesthetics moved away from an empiric of bodily sensations and towards discussions of taste. *Aisthesis* has classically been subsumed to the "higher" philosophical values of rationality and logic, as if a critical mind cannot possess a sensing body. Many philosophers have displayed forms of preferential treatment for the ideal of a calculating brain "unimpeded" by emotionality. According to Susan Buck-Morss (1992), this denial of embodied sensation had its corollary in advances in medicine: she traces a steady incline in the invention and prescription of anaesthetic drugs in the nineteenth century, roughly analogous to the transformation of aesthetics from a philosophical category interested in embodied experience to one about artistic judgement. The use of anaesthetic medicine went so far that it was routinely prescribed as a sleeping aid and even used on children by working parents. Anaesthetics became the most common way to alleviate the effects of overwork and the general "'wear and tear' of modern life" (Buck-Morss, 1992: 19).

But Buck-Morss goes even further: "Beginning in the nineteenth century, a narcotic was made out of reality itself" (1992: 22). She is lamenting a flooding of the senses by environmental stimuli—phantasmagoria. The constant overstimulation through the display of commodities and their substitutes and signifiers is a distraction of the senses and ultimately a form of social control. The term phantasmagoria harks back to the early nineteenth century when it referred to theatrical artefacts that showed frightening images and optical illusions (Jennings, 2006: 13). Following Marx's use of the term, it was adapted by Walter Benjamin to describe the role commodities play in social experience (Cohen, 1989: 88). For Benjamin (1991), cities were particularly advanced sites of phantasmagorical overstimulation and he talks specifically about "the phantasmagoria of Parisian life" (p. 543) when meditating on the Parisian arcades and their boundless display of consumables. More and more, people are exposed to sensory overload, an overload later described as "spectacle" (Debord, 1994). Benjamin (1991) equates the shock felt in the urban environment to the desensitised experience of a factory worker (p. 632) and thus reveals the bridge he saw between urban experience and alienating labour exploitation.

Georg Simmel, too, feels a sense of alienation inherent to the modern city. In his widely cited essay "The Metropolis and Mental Life", Simmel (2006) argues that cities had become centres of commercial transactions and as a result transformed their subjects' experiences and behaviours: a cultural change bound urban "spirit" more and more to its manifest environment and away from the individually sensing and imagining subject. The city's rapidly and constantly changing stimulations made it impossible to register and emotionally react to all of them adequately and, as a result, the modern mind became more and more calculating in this "rampant growth [*Überwuchern*] of objective culture" (Simmel, 2006: 39). The new objective culture on the rise in the metropolis links alienation of personal experience directly to contemporary labour, according to Simmel. In a polemic tone, he mourns how the division of labour reduces the personality of a subject by requiring of them an increasingly one-dimensional behaviour. For these thinkers, the city serves as a place and object to think about the impacts of modern capitalism. As drawn out by Timothy Mitchell (2002), the metropolis was a way for Simmel and many of his contemporaries to think about the accelerating and densifying processes of transaction that would later occupy debates about "the economy". Simmel and Benjamin were both interested in the effects of economic and labour relations on the city, as well as its subjects, and the desensitisation—almost disembodiment—they felt was caused by these relations.

According to them, alienation occurs both in the workplace and outside of it, and is directly related to the separation of the lived world from the subject's bodily experience. This link between contemporary labour conditions and a reduction in affective capacity is shared by Franco "Bifo" Berardi, who diagnoses a "widespread condition of anaesthesia" (Berardi, 2017: 53). "Being 'cheated out of experience' has become the general state", Buck-Morss summarises, and "its goal is to *numb* the organism, to deaden the senses, to repress memory" (Buck-Morss, 1992: 18). This focus on sensation in Benjamin's writing shows, according to Buck-Morss, an understanding of alienation that is grounded in cognition rather than an economic calculation. While workers in the factory are forcibly numbed into reflexive responses to the tasks and triggers of machinery, people outside are flooded by an overstimulation of superficial experience. Both dynamics alienate sensing subjects from their surroundings. Alienation as an aesthetic category, if we follow these thinkers, is the negation or coerced mediation of embodied experience. This is particularly traumatic since our nervous system "is not contained within the body's limits" but passes through both subject and environment, forming what Buck-Morss calls the "synaesthetic system" (Buck-Morss, 1992: 12–13). Doing away with the dualistic separation between inside and outside endorses the perceived relation between external conditions and sensory capacity and helps embrace a notion of the city as a synaesthetic system.

Benjamin and Simmel have received well-founded scrutiny for eulogising an imagined authenticity of bygone urban existence (Amin and Thrift, 2002: 32–39). Their argument largely relies on a nostalgic humanism in which the city is diagnosed as losing its former greatness, increasingly restricting its citizens' experience and everyday life. Nonetheless, the connection between a shared economic context and individual experience has been further developed to great effect by geographers and other scholars, as the following section draws out.

Aesthetics and Affect in Cultural Geography

A growing literature of critical geography is paying attention to embodied qualities of economic life (Thrift, 2004; Christie et al., 2006; Stewart, 2007; Anderson, 2016; Hitchen, 2021; Bissell, 2022). By investigating how neoliberalism shapes the bodies of its labouring subjects in their capacity to "affect and be affected" (Anderson, 2016: 735), cultural geographers are shifting debates on economic activity towards their affective and aesthetic dimensions. Various strands of affect and aesthetics have enabled geographers to advance their understandings of spatial "atmospheres" (Hawkins and Straughan, 2015, p. 291) and of shared "structures of feeling" (Williams, 1977: 128). Attention to these dimensions helps to engage the complex embodied elements of everyday life and produce subject-centred investigations. For instance, Anderson et al. (2020) highlight the "multiple affective and emotional worlds through which people live capitalisms" (p. 422) by researching users of digitally mediated payday loans in the United Kingdom and their resulting forms of living. Richardson (2015) underlines how the early years of AirBnB produced a sharing economy that capitalised on affect as hosts share their actual homes. In research about female millennials in Canada, Worth (2016) argues how a shared feeling of precarity is produced and how a sense of interdependence can ultimately foster worker agency. Cockayne (2016) argues that affect plays a critical role in how workers passionately attach and consolidate themselves to precarious working conditions. Bissell's (2022) research looks at how forms of work limit and transform people's capacity to be affected and how a disassociation is at times a protective mechanism, thereby highlighting the ambivalence and complexity of embodied dimensions of labour.

This reflects a development of interests from how labour affects workers' capacities to sense and feel. At stake are not only subjective feelings but also social relationality produced through shared context. Anderson (2016) invokes Raymond Williams' concept of structures of feeling to argue how neoliberal atmospheres enter everyday life: structures of feeling describe "forming and formative processes" that "exert palpable pressures and set effective limits on experience and on action" (pp. 745–746). What is identified is a structure of desensitisation, similar to what Buck-Morss calls anaesthetics. One collective feeling subjects are left with is "cruel optimism", which describes the attachment to objects or ideas that are ultimately not in one's interest in an attempt to legitimise and live with constant precariousness (Berlant, 2011).

The shift in attention to the aesthetic dimensions of sensing and feeling, of being affected, often draws out how subjects are alienated from their aesthetic and sensory capacities, echoing earlier discussions by theorists such as Simmel, Benjamin, and Buck-Morss. Bissell's (2022) work on "the anaesthetic politics of being unaffected" directly references this critical tradition as a viewpoint he labels "constraining anaesthetics" (pp. 89–90) before moving, with the help of Berlant's cruel optimism and Katz's typology of resilience, reworking, and resistance, towards an enabling sense of anaesthetics in which subjects "unfeeling" is used as a protective mechanism in the face of exploitative, alienating, and precarious labour relations (Bissell, 2022: 90–92). By highlighting how a reduced capacity to feel may at times be beneficial, Bissell is able to highlight the ambiguous and complex interrelations between aesthetics and alienation.

Although much of the literature—Bissell a notable exception—does not explicitly work through a notion of aesthetics, the connections in content are manifold. Despite the fact that affect enjoys a variety of non-exclusive meanings and uses, tied more or less stringently to differing philosophical traditions, it is usually accepted that affect is concerned with an embodied and relational account of the world (Anderson, 2013: 456–458) and in this sense shares a fundamental resemblance to the original aesthetic concerns of *aisthesis*. Dixon and Straughan (2013) suggest that affect and its focus on embodied forms of knowing are inherently bound up with aesthetics. The relationality inherent to the concept and its emphasis on the affective composition of bodies in their lifeworld bears a clear resemblance to Buck-Morss' concept of a synaesthetic system in which an embodied perception of the world cannot be separated from the world that is being perceived (Buck-Morss, 1992: 11–13).

Recent interest in embodiment and perception as crucial factors in contemporary labour geography highlights the fruitfulness of an account of the aesthetic grounded in *aisthesis*. Despite the current preference for a terminology of affect, the above appraisal of Simmel, Benjamin, and Buck-Morss can contribute to much of this literature on subjects' capacities to feel and perceive. The growing sense that both urban experience and labour relations benefit from geographical accounts that centre such embodied capacities make an "aisthetic" lens timely and constructive.

Bicycle Messengers and Their Aesthetic Lifeworld: Two Vignettes

Earlier, this chapter suggested that bicycle messengers often express a sense of contentment and identity that is superficially at odds with the precarity of their work and that aesthetics may be a way of understanding why and how this resilience occurs. Over the past few years, across interviews and field observations, this multidimensional, often positive assessment of their own situation struck me as a recurring theme. It is not uncommon for couriers to express pride in their work and the skills and knowledge it enabled them to acquire, with the physical elements of the job, as well as a kind of unmediated connection to the city around them often given as an explanation. To illustrate these themes, two vignettes are offered below that are taken from semi-formal interviews conducted between June and September 2022 with participants in

Glasgow and Zürich. They were selected from interviews conducted in a number of cities, also including Los Angeles, Seattle, and Bremen. In addition, between 2014 and 2022 I have worked both full-time and part-time in Glasgow, Melbourne, Helsinki, and Los Angeles. This has allowed me to engage in in-depth work-based ethnography and has informed both my experiences and analysis.

One of the interviews was conducted virtually while the other was a ride-along interview where I was following the messenger for the entirety of his workday. This technique has proven particularly fruitful since it allows for an interview enriched by the embodied experience of the labour process; the data was gathered in situ, allowing both place and work to inform my reception and in turn to shape my questions depending on real-life situations of the courier's life-world. Following Casey, Anderson (2004) uses the notion of "constitutive co-ingredience" of people and place to develop the benefits of having "conversations in place" (pp. 255–256). In particular, Anderson argues that talking whilst walking has the potential to harness place in order to access a deeper insight into the socio-spatial context of human knowledge construction of their lifeworld. Similar to these go-along interviews, ride-along interviews with messengers and couriers are a productive way to harness place and produce insights that situate the subject's knowledge within the context of the city they are so intimately bound to. Adding a layer of embodied experience enables a deeper understanding on the side of the researcher. Another advantage in this particular case is that by producing a form of workplace ethnography in which labour processes themselves are built into the interview, the researcher is able to understand the courier's knowledge of the world through a direct account of their experiences. While work-based ethnography already has a well-supported tradition when it comes to researching delivery riders (Fincham, 2006; Kidder, 2006; Popan, 2021), ride-along interviews offer a valuable addition to the ethnographic toolkit of workplace geographies (Elwood and Martin, 2000; Brannan et al., 2007; Hodson, 2008), with a focus on embodied research. Considering the topical relevance of sensory experience and embodied everyday practices, and as feminist geographers have also emphasised for decades, methods that centre on lived experience and conducting research "through the body" and with attention to the issues of embodiment seem not only relevant but a necessary inclusion (Sheller and Urry, 2006; Büscher and Urry, 2009; McMorran, 2012; van Doorn, 2017).

Noé

Noé has chestnut eyes and hair of the same colour curling out from underneath his helmet. He speaks German with a gentle and melodic Swiss accent, sprinkled with exclamations in Spanish. Noé's parents are from Spain, but he grew up in Bern before moving to Zürich for his studies. He is 27 years old. Noé began working as a messenger while completing a Master's degree in Environmental Sciences and has been working at his current company for the past two years. Before that, he worked for a small collective in Zürich that delivers fresh fruit and vegetables to their subscribers on cargo bikes. When I ride along during his shift it is one of the colder days in June, just warm enough to ride in shorts and a jersey. The sky is overcast and a slight northeasterly breeze, a Föhn wind, ripples Lake Zürich. Whenever traffic allows it, I ride up next to him to ask questions and listen to his responses. Each time we stop for a pick-up or delivery, we continue the conversation, and I take notes. Noé likes to ride fast and more than once I have a hard time staying on his wheel. He is hunched over his handlebars in an aerodynamic position, despite the massive yellow backpack that matches the colour of his bicycle. After one pick-up in the morning that takes us to his company's headquarters, we get assigned a number of locations along the western shore of Lake Zurich. "We can really go full throttle", he beams at me with a smile that will not leave his face for the next few hours.

It is obvious that Noé enjoys his job and that what he likes most about it is riding his bike. He takes on the rest of his workday in a playfully competitive manner, always pushing to get everywhere as quickly as possible, no matter if a deadline requires it. We leave Zürich and head south on Seestrasse, occasionally turning off and climbing to nearby addresses for a pick-up. Noé only checks his phone for directions on two occasions and we only miss an address by a block once. As we turn around, Noé curses his mistake, in jest admitting to a great embarrassment: "I almost turned there—no idea why I went past". Similar to London's black cab drivers, who refer to their almost complete ability to select the best route without consulting a map as "the knowledge", messengers take pride in the experience they build navigating their city. This embodied knowledge of their surrounding city is performed both on a macro and a micro level: Noé knows how to ride to any of today's clients, crossing Zürich multiple times on runs that reach from the ETH campus in the north to villages as far south as Horgen and Meilen. On each of these runs, he understands his own abilities and paces himself correctly. Meanwhile, the same embodied knowledge becomes visible on a small scale, as he reads and moves through intersections with ease and fluidity, skipping over tram tracks and snaking past stop-and-go traffic.

On our way to the ETH campus, we have to climb the slopes of the Käferberg. Laughing, Noé says, "We could take the bus, too. Maybe it's faster", before accelerating out of the saddle and pushing on. He says that unlike some of his colleagues he enjoys assignments that leave behind the city centre; he likes the challenge and the way his body feels after a hard day of intense work. We have lunch on campus, where I ask Noé about choosing to work as a messenger after studying at a prestigious university. "I moved here for my studies and started to work as a messenger as I was finishing. It suits me: I get to go everywhere in the city, I get paid to ride". Noé and his housemate Jordi, who works for the same company, like that they get to ride bikes for a living, and that they are not stuck in a full-time job. For their holidays, the two are planning a bike tour through the Pyrenees, he tells me. "It's gonna be mega nice", he smiles. When I ask him what he likes most about the profession, he vaguely nods his head to his surroundings and says "all of this; riding".

Noé looks at his most happy when on top of his bicycle. He tells me of the fun of "playing in traffic", an engagement with the city around him that requires experience and skill, which Kidder has also described as "deep play" when researching riders in Seattle and San Diego (Kidder, 2017). Even when we board a ferry to cross Lake Zurich to its eastern shore for a pickup at a doctor's office, Noé remains glued to his seat. He spends almost the entire passage balanced on his bike, hands off the handlebars and stretched out for stabilisation, one foot on the pedal, the other one on his front wheel to push it forth and back in little movements, keeping himself upright. "Trackstanding" is a vital skill for many messengers and is often performed at red lights to avoid a full stop. At Cycle Messenger World Championships and other events, riders even compete in trackstand competitions. After work, Noé, Jordi, and other colleagues meet in a small square close to their office for beers and even there, they are pulled back onto their machines every few minutes. Forgotten are the tired legs, after long miles and strenuous climbs, as the messengers, with child-like excitement jump onto their steeds to practice trackstands and backward circles or play rounds of "foot-down", where riders attempt to force others to put a foot on the ground by cutting them off or bumping against them. Noé and his colleagues are content to be messengers and enjoy the insight it grants them to their city. They are proud of their abilities on a bike and approach their job with playfulness and pleasure.

Mark Paterson (2015), in an investigation of the aesthetics of moving bodies, draws out a history of kinaesthetic enjoyment, or "the pleasures of movement" (p. 36). Drawing from phenomenological and medical literatures of the eighteenth and nineteenth century before turning to contemporary dance and performance, he reconstructs a far-reaching understanding of kinaesthetic pleasures and the particularly meaningful capacity for collective experience of movement (Paterson, 2015: 46). This seems to be a key ingredient in the meaning and identity Noé and others draw from their work and the collective and personal sense of joy I witnessed that day, and might be one reason why one of the company's riders has recently told a local newspaper he would like to work as a bike messenger until retirement (Grundlehner, 2019).

James

James is in his early 50s. His grey hair and beard have a red tint. When we meet across time zones, his weather-beaten face is wearing a warm smile. He looks over the frame of his glasses at the screen, not at the camera but just below it, at my face, I imagine. James sits in the living room of his shared flat in Govanhill, a neighbourhood on Glasgow's southside. He is wearing a bunnet, the tweed flat cap often worn by working-class men in Scotland, and a black t-shirt with the logo of a London courier company. Throughout the interview, I feel a sense of tenderness when he talks about his hometown. The Glaswegian has been a bicycle messenger for almost a quarter of a century.

When I ask James to describe the work of a bike messenger, his immediate response highlights his sensory and embodied experience:

> Up early, get ready, coffee, you know... Waiting. Waiting for your first call. [The] first ride of the morning is always quite exhilarating, and my eyes always stream of water. It's as if I am crying for the first ten minutes of the day. It doesn't matter what condition I am in, when I get to my first pick-up, I am already fully awake and alert and alive.

Asked about his job, James' reaction is to describe a phenomenological experience. He highlights rich, sensory qualities before moving on to describing his connection to the city. By now, he knows the traffic so intimately that he rarely has to stop. "It's probably where I am at my most comfortable. And the busier the better, obviously". This and similar engagements with the city are common among messengers who often express admiration for how skilled a rider can move through traffic and how much his or her own rhythm matches the rhythms around them. James likens the state of riding as meditative; he feels a sense of being in the moment, constantly present, a state from which he draws comfort. Experience has taught him to be ready for anything while riding and the flow he enters enables James to detach from other thoughts or worries. This, too, is common for delivery riders. Many liken riding through traffic to a profoundly fulfilling state of flow (Kidder, 2017).

With the pandemic, a lot of his former clients have fallen away and in order to supplement his income he started to do food deliveries. "I find the food delivery great fun as well. If you are juggling three or four of those apps at a time, it can be real good fun". This tactic known as multi-apping (Popan, 2022) is common among food delivery riders whose low wages and precarious working conditions often force them into logging onto multiple platforms at once to make a living wage. But while this is usually correctly decried as a warning sign of the exploitative nature of gig work, James approaches it with positivity. He welcomes the added complexity and challenge of completing multiple orders at once and says that the physical and mental challenges of his profession are what keep him engaged. He likes that his job requires quick thinking and for him to stay in good physical shape. To him, there is no inherent difference between the documents he delivered two decades ago and the kind of cargo he is moving now; what defines the work for him is riding his bike, responding to the traffic around him, and being a part of the city.

It is obvious that James identifies greatly with being a courier. He tells me about the international community and sense of solidarity, how friends organised races to raise funds for him when he got injured. By participating in European and World Championships, as well as visiting friends he met at these events, the profession encouraged him to travel internationally, as well as host friends from abroad. When I ask him about the importance of the job in his life, he nods. "It's massive", he says, "it's massive". He tells me with pride about how many people know and recognise him as he haunts the streets. Other delivery riders, many from northern Africa and eastern Europe, call him uncle. Cab and bus drivers, even pedestrians, greet him. He tells me about the Glasgow bin lorry crash in 2014 where he was a first responder, helping a woman he later found out shared his last name. Being a messenger, he says, allowed him to become a participant in his city, rather than a bystander.

> If I go along the river [I can] sit with asylum seekers and have a beer with them, or I can go to a fancy bar in the West End and sit with the hoi polloi. It's like you are almost a conduit between them all, you know.

James is describing an intimate sense of presence in the city, a sort of access to its rhythms. The city seems to form a part of James' synaesthetic system, rather than something wholly external.

It is clear that being a messenger has a formative influence on how he sees himself and the world around him. When asked about his time before being a bicycle courier, he responds: "I didn't start growing as a person until this job". He says he does not recognise the person he used to be. He says that he could have chosen a different path in life if money was a priority but makes clear that he enjoys being a courier, the embodied nature of his daily practice, as well as the access it has given him to the city he lives in. When I ask about his age and future, he makes clear that he would like to stay in good enough shape to keep doing his job, and he is confident he has an advantage: "I ride my bike in traffic as well as anybody". This personal pride in his work, as well as his appreciation of the nuanced skills involved in the work, matches studies of craft and craftworkers. As Randy Hodson (2001) describes, "a defining characteristic of craft workers is intense pride in their work… The personal identities of craft workers thus tend to be firmly grounded in their occupational skills and activities" (p. 142). Indeed, for many craftworkers, their capacity to do good work is their "primordial mark of identity" (Sennett, 2008, 25). Despite the working conditions, poor pay, physical dangers, and minimal requirements for the profession, bicycle messengers, at least those classifiable as lifestyle messengers, seem to more or less express the traits of craftworkers.

Aesthetics as Minor Theory

Both of these vignettes can serve as examples of messengers' construction of themselves and their lifeworlds. Despite an otherwise alienating labour that is dangerous and underpaid, messengers identify greatly with their work and partially draw from it both a sense of self and a sense of the city around them. Aesthetics helps to explain: (i) how the practice of working as a courier can afford a sensory experience of the city and oneself unavailable to many and (ii) why messengers' capacities to feel and produce embodied knowledges is an important source of meaning and enjoyment. But of course, this is not an exclusive theory that claims complete explanatory power. Rather, it comes closer to what Vattimo (2012) calls weak thought and what Katz (1996) refers to as minor theory, which can be thought alongside other explanations and theories. After all, this aesthetic capacity is not the only cause of workers' self-worth, and neither does embodied experience always led to contentment. Some couriers do not display the same pride or excitement about their work. Other elements alongside it, such as legal status, or the embeddedness in a local culture and community, are equally important factors that contribute to higher levels of identification possible for messengers. Notably, there also seems to be a general drop-off in levels of identity and self-valorisation between "classic" bicycle messengers and food delivery workers that is likely connected to different levels of precarity, subcultural integration, and social integration in general.

Despite these relativisations, an aesthetic lens still makes a meaningful contribution and is able to account for many of the reasons couriers routinely give to explain their investment in the job: It explains in part the deep-rooted enjoyment couriers find in their work that helps them elevate it beyond "mere labour" and cope with the exploitative working conditions they are facing. Furthermore, by raising aesthetics as a minor theory, this chapter offers a non-exclusive explanation.

Conclusion

Many bicycle messengers display a strong sense of self-valorisation, pride, and enjoyment in relation to their profession. Despite the hardships and precarity of a work that is unstable, underpaid, and dangerous, they identify greatly with their job. This is aided by and simultaneously fosters a strong subculture and community, complete with races and even international events and competitions that celebrate their work and practice. While this is superficially at odds with the alienating and exploitative nature of work, many messengers have a positive relationship with their work that is different from "cruel optimism" (Berlant, 2011). Rather, following a critical tradition of aesthetic theory which revisits the original meaning of *aisthesis* as sensory experience and highlights contemporary desensitisation in a process of anaesthetics, this chapter has argued for treating aesthetics as an explanation for messengers' continual capacity to "make sense" of themselves and the city around them through embodied experience.

As aesthetics is retraced to sensory capacity, this critical tradition proposes anaesthetics as the reduction of affective capacity. The production of anaesthetic experiences is an essential feature of the modern city according to thinkers such as Simmel and Benjamin. Contemporary scholars have returned to their provocations in their concern for the affective and embodied dimensions of the capitalist economy. In this lineage, the aesthetic dimension in the work of messengers reveals their richly affective sense to the city—and reciprocally themselves—guided by everyday embodied interaction that generates a sense of participation and contribution. This chapter does not claim that aesthetics is an exclusive explanation of this complicated relationship nor does it claim that this aesthetic capacity is exclusive to bicycle messengers. Rather, revisiting aisthesis as minor theory is suggested as a useful lens among others for couriers' comparably high resilience and self-worth.

With a growing body of literature in cultural geography and related fields interested in the affective and embodied dimensions of capitalism and the atmospheres and structures it produces, and how those influence subjects' capacities to act and feel, messengers' sensory relationship to their work and the urban fabric provides an interesting case. The aesthetics inherent in their work on mobility allows couriers to construct knowledges of the world that afford them a strong sense of self, grounded in embodied, sensory experience. The vignettes presented in this chapter provide a complex picture of a precarious and exploited workforce that is able to produce meaning, joy, and an intimate sense of the city. Rather than leisurely musing the city's essence from the vantage point of a detached spectator, couriers are entangled with the contemporary city and economy, always simultaneously experiencing the world around them and their own interpellation within it. This suggests a synaesthetic experience, by which their mobility is valued as a way of experiencing and knowing the city. It also bears elements of craftwork as messengers stress the skills required to "read" the spaces and rhythms of traffic, and to build a profound and intimate knowledge of the cities they navigate.

This chapter contributes to the understanding of how messengers produce meaning and joy through the embodied practice of delivery work, an aesthetic practice in so far that it shapes their sensory experience of the world. But this is not all there is to it. The argument so far has highlighted how this aesthetic capacity allows messengers to celebrate their mobility, despite the labour conditions that underlie it. Simultaneously, however, this also legitimises the very conditions that form the basis of their exploitation. As couriers carve out aspects of their work that provide enjoyment, pride, and community, they find ways to accept what else the practice comes with: poor pay, precarious conditions, and high personal risk. The kinaesthetic enjoyment of their work, the sense of connection it offers them to their city, and the subculture that comes with it, all become reasons why delivery work becomes more acceptable. What offers resilience on the one hand is therefore also in the interest of companies who are able to capitalise on a happy workforce.

With platforms already routinely building gamification into their algorithmic management strategies (van Doorn and Chen, 2021), elements such as enjoyment, embodiment, and playfulness are increasingly instrumentalised in how employers maximise efficiency and profitability. As such, companies either do not mind if workers find ways to actively tease out more meaning and aesthetic value from their work, or they actively encourage it. Many smaller companies with ties to the subcultural space openly embrace, even sponsor, events such as championships or "alleycats". And even large start-ups at times capitalise on the subculture: as one informant told me, when he applied to work for an app-based delivery service in Berlin, the final stage of the interview was a race against other applicants to an address in the city. This was supposed to test their skills and ability on the bike, but of course flirted with a subculture that has long simulated their work in community-run events. The start-up purposefully incorporated the practices of play through which messengers celebrate their work while exposing potential subcontractors to unnecessary risk as they competed against one another for a chance at employment.

Companies can actively instrumentalise the aesthetic dimensions of this embodied practice, but even when it is only a by-product in which a messenger finds joy, unencouraged by the employer, companies still benefit from it. As such, it would be insufficient to call the aesthetic capacity couriers produce a subversive mobility. But simultaneously, this embodied capacity for kinaesthetic enjoyment is not entirely recuperated either. It does provide a sense of identity and resilience that is not fully containable. As such, the aesthetic capacity produces somewhat of an oblique space: neither in absolute opposition to nor fully in line with the conditions of their labour, it assumes its meaning "between surfaces" (Foucault, 1980: 169). In that it exceeds to be neatly contained by either, this aesthetic of delivery work holds both opportunity and risk. It is the task of its practitioners to make sure it spills not into platform-controlled gamification, but into what Henri Lefebvre (1996) enigmatically called the *fête*: the restitution of the city as a space of festivity and leisure through a transformation of everyday life.

References

Amin, A. and Thrift, N. (2002). *Cities: Reimagining the Urban*. Polity.

Anderson, B. (2013). Affect and Emotion. In *The Wiley-Blackwell Companion to Cultural Geography*, edited by N. C. Johnson, R. H. Schein, and J. Winders. Wiley-Blackwell, pp. 452–464.

Anderson, B. (2016). Neoliberal affects. *Progress in Human Geography*, 40(6), 734–753.

Anderson, B., Langley, P., Ash, J., and Gordon, R. (2020). Affective life and cultural economy: Payday loans and the everyday space-times of credit-debt in the UK. *Transactions of the Institute of British Geographers*, 45(2), 420–433.

Anderson, J. (2004). Talking whilst walking: A geographical archaeology of knowledge. *Area*, 36(3), 254–261.

Barns, S. (2019). Negotiating the platform pivot: From participatory digital ecosystems to infrastructures of everyday life. *Geography Compass*, 13(9), e12464.

Benjamin, W. (1991). *Gesammelte Schriften I*, edited by R. Tiedemann and H. Schweppenhäuser. Suhrkamp.

Berardi, F. (2017). *Futurability: The Age of Impotence and the Horizon of Possibility*. Verso.

Berlant, L. G. (2011). *Cruel Optimism*. Duke University Press.

Bissell, D. (2022). The anaesthetic politics of being unaffected: Embodying insecure digital platform labour. *Antipode*, 54(1), 86–105.

Brannan, M., Pearson, G., and Worthington, F. (2007). Ethnographies of work and the work of ethnography. *Ethnography*, 8(4), 395–402.

Buck-Morss, S. (1992). Aesthetics and anaesthetics: Walter Benjamin's artwork essay reconsidered. *October*, 62, 3–41.

Büscher, M. and Urry, J. (2009). Mobile methods and the empirical. *European Journal of Social Theory*, 12(1), 99–116.

Caprotti, F., Chang, C., and Joss, S. (2022). Beyond the smart city: A typology of platform urbanism. *Urban Transformations*, 4(4), 1–21.
Christie, J., Gough, R., and Watt, D. (2006). *A Performance Cosmology: Testimony from the Future, Evidence of the Past*. Routledge.
Cockayne, D. G. (2016). Entrepreneurial affect: Attachment to work practice in San Francisco's digital media sector. *Environment and Planning D: Society and Space*, 34(3), 456–473.
Cohen, M. (1989). Walter Benjamin's phantasmagoria. *New German Critique*, 48, 87–107.
Day, J. (2015). *Cyclogeography: Journeys of a London Bicycle Courier*. Notting Hill Editions.
Debord, G. (1994). *The Society of the Spectacle*. Zone Books.
Dixon, D. and Straughan, E. R. (2013). Affect. In *The Wiley-Blackwell Companion to Cultural Geography*, edited by N. C. Johnson, R. H. Schein, and J. Winders. Wiley-Blackwell, pp. 36–38.
Dizard, W. P. (2001). *Digital Diplomacy: U.S. Foreign Policy in the Information Age*. Praeger.
Downey, G. J. (2002). *Telegraph Messenger Boys: Labor, Technology, and Geography, 1850–1950*. Routledge.
Eagleton, T. (1990). *The Ideology of the Aesthetic*. Blackwell.
Elwood, S. A. and Martin, D. G. (2000). "Placing" interviews: Location and scales of power in qualitative research. *The Professional Geographer*, 52(4), 649–657.
Fincham, B. (2006). Back to the "old school": Bicycle messengers, employment and ethnography. *Qualitative Research*, 6(2), 187–205.
Fincham, B. (2007). 'Generally speaking people are in it for the cycling and the beer': Bicycle couriers, subculture and enjoyment. *The Sociological Review*, 55(2), 189–202.
Foucault, M. (1980). Theatrum philosophicum. In *Language, Counter-Memory, Practice*, edited by D. Bouchard, Cornell University Press. pp. 165–196.
Grundlehner, W. (2019, August 14). Ich möchte als Velokurier pensioniert werden. *Neue Zürcher Zeitung*. https://www.nzz.ch/finanzen/ich-moechte-als-velokurier-pensioniert-werden-ld.1498025
Harvey, D. (2010). *A Companion to Marx's Capital*. Verso.
Hawkins, H. and Straughan, E. R. (Eds.). (2015). *Geographical Aesthetics: Imagining Space, Staging Encounters*. Ashgate.
Hitchen, E. (2021). The affective life of austerity: Uncanny atmospheres and paranoid temporalities. *Social & Cultural Geography*, 22(3), 295–318.
Hodson, R. (2001). *Dignity at Work*. Cambridge University Press.
Hodson, R. (2008). The ethnographic contribution to understanding co-worker relations. *British Journal of Industrial Relations*, 46(1), 169–192.
Jennings, M. W. (2006). Introduction. In *The Writer of Modern Life: Essays on Charles Baudelaire*, edited by W. Benjamin. Harvard University Press, pp. 1–25.
Katz, C. (1996). Towards minor theory. *Environment and Planning D: Society and Space*, 14(4), 487–499.
Kidder, J. L. (2006). "It's the job that I love": Bike messengers and edgework. *Sociological Forum*, 21(1), 31–54.
Kidder, J. L. (2017). *Urban Flow: Bike Messengers and the City*. Cornell University Press.
Lefebvre, H. (1996). The right to the city. In *H. Lefebvre: Writings on Cities*, edited by E. Kofinan and E. Lebas, (pp. 63–184). Blackwell.
Marx, K. and Engels, F. (2008). *Werke. 23: Das Kapital, Bd. 1*. Dietz.
McMorran, C. (2012). Practising workplace geographies: Embodied labour as method in human geography. *Area*, 44(4), 489–495
Mendelsund, P. (2021). *The Delivery*. Picador.

Mitchell, T. (2002). *Rule of Experts: Egypt, Techno-politics, Modernity*. University of California Press.
Orr, W., Henne, K., Lee, A., Harb, J. I., and Carneiro Alphonso, F. (2023). Necrocapitalism in the gig economy: The case of platform food couriers in Australia. *Antipode*, 55(1), 200–221.
Paterson, M. (2015). On *aisthêsis*, 'inner touch' and the aesthetics of the moving body. In *Geographical Aesthetics: Imagining Space, Staging Encounters*, edited by H. Hawkins and E. R. Straughan. Ashgate, pp. 35–51.
Popan, C. (2021). Embodied precariat and digital control in the "gig economy": The mobile labor of food delivery workers. *Journal of Urban Technology*, 33(1), 109–128.
Popan, C. (2022, July 28). Multi-apping in the gig economy: Between entrepreneurship and a race to the bottom. In *Transforming economies and essential services: Delivery workers and the new capitalist offensive* [Panel Presentation] EASA 2022: Transformation, edited by L. Steur and K. Monroe (Chairs), Hope and the Commons, Belfast.
Richardson, L. (2015). Performing the sharing economy. *Geoforum*, 67, 121–129.
Sadowski, J. (2020). Cyberspace and cityscapes: On the emergence of platform urbanism. *Urban Geography*, 41(3), 448–452.
Sennett, R. (2008). *The Craftsman*. Yale University Press.
Sheller, M. and Urry, J. (2006). The new mobilities paradigm. *Environment and Planning A: Economy and Space*, 38(2), 207–226.
Simmel, G. (2006). *Die Großstädte und das Geistesleben*. Suhrkamp.
Srnicek, N. (2017). *Platform Capitalism*. Polity.
Stewart, K. (2007). *Ordinary Affects*. Duke University Press.
Thrift, N. (2004). Intensities of feeling: Towards a spatial politics of affect. *Geografiska Annaler: Series B, Human Geography*, 86(1), 57–78.
van Doorn, N. (2017). Platform labor: On the gendered and racialized exploitation of low-income service work in the "on-demand" economy. *Information, Communication & Society*, 20(6), 898–914.
van Doorn, N. and Chen, J. Y. (2021). Odds stacked against workers: Datafied gamification on Chinese and American food delivery platforms. *Information, Communication & Society*, 20(6), 898–914.
Vattimo, G. (2012). *Weak Thought*, edited by P. A. Rovatti and P. Carravetta. SUNY Press.
Widera, J. (2021). *Sign and Return*. Verlag für Handbücher.
Williams, R. (1977). *Marxism and Literature*. Oxford University Press.
Woodcock, J. and Graham, M. (2020). *The Gig Economy: A Critical Introduction*. Polity.
Worth, N. (2016). Feeling precarious: Millennial women and work. *Environment and Planning D: Society and Space*, 34(4), 601–616.

5 Writing the City Clean

Sketching Conflictual Aesthetics of Street Art in Linz, Austria

Friederike Landau-Donnelly and Stefanie Fridrik

Introduction: Conflictual Aesthetics in and of the City

Drawing on empirical insights from our case study of street art and graffiti governance in the post-industrial city of Linz, Austria, this chapter argues that cleanliness can be problematically enforced through street art and graffiti policies to advance a specific notion of aesthetics in the city. While urban and geographic scholarship has developed theories and concepts of aesthetics in urban contexts (see Ghertner, 2010, 2015; Dikeç, 2015; Hawkins and Straughan, 2015), urban cultural research continues to struggle with the ambivalent terms of aesthetics and art both in and of the city (Luger and Ren, 2017). To approach the political underpinnings of the aesthetics in and of the city, Asher Ghertner (2010: 202) provides a Foucault-inspired account of the "aesthetic ordering of the city" as a specific governing technique, "in which the legality and essential features of space can be determined entirely from a distance and without requiring accurate survey or assessment". What Ghertner also terms aesthetic governmentality is steered by acclaimed experts, planners, and other public–private authorities. Similarly, our understanding of the aesthetic politics of the city resonates with Millie's (2016) Rancière-inflected understanding of aesthetic orders as inherently political. Millie (2016: 4) considers urban aesthetic orders to actualise when "the city is tamed and beauty is found in an ordered urbanity". Informed by these debates, and with regard to the aesthetics *in* the city, we wonder about the manifold affective and (im)material connections amongst humans and more-than-humans in relation to aesthetic expressions that can manifest in endless ways (on walls, through paint, on and across built structures, via plants, weeds, in ruins etc.). How are these connections influenced by cultural and other urban policies? What role do different aesthetic expressions play in the creative making of the city and urban placemaking? How do they further relate to urban (re)branding or policy idea(l)s such as cleanliness and order?

In this chapter, and closely related to the aesthetics *of* the city, we are curious about the conditions of (im)possibility that produce such aesthetics of the city with a particular focus on its political underpinnings—what we call an *aesthetic politics of the city*. In other words, we are focusing on the contingency

DOI: 10.4324/9781003335993-5

of what kind of aesthetics of the city are seemingly realisable in a given order (Landau-Donnelly, Schad-Spindler, and Fridrik, 2023). While aesthetics might be even more tricky to unpack conceptually than public art, we push for an open-ended engagement with aesthetics that considers the creative, artistic, and political forms of collective (or individual) meaning-making about what art "is" (and what it isn't). Wrestling with aesthetic politics, too, includes questioning whose opinion about art and aesthetics matters (and whose doesn't) and who is even capable of creating art (and who assumed not to be). In this sense, aesthetics blend epistemologies of the recognition and legitimation of art that can sediment into forms of control, governmentality, or what Rancière (2010) calls "the police".

Alongside Rancière, we mobilise Marchart's (2019) term *conflictual aesthetics*, which sets up our understanding of aesthetics as inherently, and irrevocably, conflictual and contested. With help of the novel term *aesthetic politics of the city*, we sketch the parameters that condition the perception, validation, and marketisation of art in public space. As arts and culture have been variously deployed in urban development, regeneration, and city-building projects (Sacco et al. 2013; Markusen and Gadwa, 2010; Miles and Paddison, 2005: McLean, 2014), it is worth studying the spatial implications of aesthetic assumptions in urban settings. Drawing on interviews with artists, cultural organisers, policymakers, and politicians in Linz, we unpack some of the tensions between the visibility of public art, which is implemented and directed by city authorities, and the responses from independent cultural and street art actors, articulating their own interpretations of aesthetic politics of the city.

Our analytical observations stem from a research project (2020–2022) on conflictual cultural politics in Austria's three biggest cities (Landau-Donnelly et al., 2023; Schad-Spindler et al., 2023). In Linz, we conducted eight semi-structured interviews with local cultural administrators, tourism managers, cultural organisers, curators, artists, and researchers. In addition, we reviewed local media reports and current cultural policy, urban planning, and tourism development documents that problematised the marketisation of street art, the contentious presence of graffiti, and the cultural-economic development of the city's harbour area. Lastly, we created situational as well as social worlds and arena maps (see Schad-Spindler, Fridrik, and Landau-Donnelly, 2023), identifying and ordering clusters of discursive constructions, individual and collective human actors, key sensitising concepts, political and economic elements, and non-human actants, which are entities that act or to whom agency is granted. Such actants also include implicated and silent actors and actants. In doing so, we analysed Linz's actorscape of street art to delineate the latently conflictual dynamics of appropriating graffiti and street art for city-making, -marking, and marketing purposes, which became subject to public critique.

In our theoretically motivated endeavour, we turn to the conflictual underbelly of aesthetics *in* and *of* the city. By mobilising conflictual theories of both space (Landau, Pohl, and Roskamm, 2021) and aesthetics (Marchart, 2019), we set out to discuss a notion of aesthetics that is permeated by *conflict* and *contingency*, and further add the aspect of *contamination*. We propose contamination as a counterpart to the notion of *cleanliness*, which we identify—in the specifically Western context of Linz—as a recurring reference point in the city's urban cultural politics and branding strategies. Notably, neither in Linz nor elsewhere, is cleanliness a neutral concept. Amongst multiple interpretations of what counts as clean, the perceived necessity and repercussions of cleanliness differ significantly—opinions and expectations about the need for, and "nature" of cleanliness can divide spouses, roommates, and neighbours, but also city governments, public and private actors, and human and more-than-human users of urban space. Think, for instance, of soil—is it "dirt" or nature's purest form of matter? In light of the ambivalent meanings of both cleanliness and contamination (the latter's most crucial function being to disturb, pollute or infest), and the politically charged implications arising from this, we examine the aesthetic consequences of Linz's objective to become the cleanest (post-industrial) city in the country. In short, in this chapter, we trace and contour the conflictual aesthetics of cleanliness as imbricated in Linz's aesthetic politics of the city.

Urban Cleanliness and the Contested Creative Making of the City

Existing conceptual approaches have sought to capture the presence of creative, artistic, and aesthetic forms in public space, ranging from terms such as public art (Cartiere and Zebracki, 2016), street art, which often subsumes graffiti (Avramidis and Tsilimpounidi, 2016; Riggle, 2010; Hoppe, 2020), or urban art (Blanché, 2015). While public art is often sanctioned to celebrate diversity, brightness, and positivity, it can also leverage negativity, pain, and trauma (Bruce, 2017; Landau-Donnelly, Forthcoming). Within the trope of street art and graffiti, its specifically *urban* character can highlight the complexity and conflictuality of urban life (Ferrell, 2016; Iveson, 2010). Graffiti as a highly contested creative form elicits a large variety of responses, ranging from appreciation and stimulation of witty and welcome commentaries in urban space, but also condemnations as vandalising, disrespectful, and illegal. As Zieleniec (2017: n.p.) argues, graffiti operates as a symbol of:

> anti-social deviant behaviour, [...] community breakdown and decline, [...] urban blight, a lack of direction, discipline and [a sign of] deviance in youth [...] graffiti presents challenges to different conceptions of what kind of city we want and also who has the power to mould, shape, adorn, decorate and control what does – and what does not – appear there, how spaces and places are organised, policed and regulated.

It is in the intellectual environment of "messy urbanism" (Chalana and Hou, 2016) that we situate reflections on the aesthetic politics of cleanliness in the Austrian city of Linz. In the attempt to counter Linz's bad reputation as a "stinking" city, caused by heavy industrial air-borne pollution during steel and chemical production, which lingered in the city up until the mid-1980s, specific notions of cleanliness have gained traction as techniques to govern public space, and within it, trajectories of (which) art is allowed therein.

Interestingly, we could not detect academic/written references to or discussions of "clean urbanism" (yet). Rather, a cursory literature review offered accounts on clean drinking water, street cleanliness, solid waste management, perceived cleanliness or smart city technologies that shall inform the latter endeavour. Reinterpreting Mary Douglas' (1966, 44) famous dictum that dirt is "matter out of place", in the context of urban art or graffiti, the question of "being out of place" concerns both the artwork that is being placed, or *dis*-placed, as well as the places that are being dirtied with art (or not). In this exploratory chapter, we play with the question of what happens when art is not *in* its officially designed place. On the one hand, this refers to the literal sense of public art's politics of location, or "lawscape", which is instructed by urban planners, zoning authorities, building/wall owners etc. (Bengtsen and Arvidsson, 2014). On the other hand, and in the broader political sense of being in or out of place, art that is not considered to be *in* its proper place might challenge societal norms or hegemonies. Examining middle-class landscapes of gated communities in Shanghai, Choon-Piew Pow (2009, 388) argues that the aestheticisation of urban spaces is entangled with neoliberal social-spatial divisions that serve as "a politics of exclusion" and urban segregation. A hegemonic aesthetic regime is consequently instated to (re)affirm class identities that are non-threatening, non-revolting, and in that sense "clean" or "proper". In the context of conflictual street or public art, this chapter focuses on graffiti as one specific creative medium to conceptualise the conflictual aesthetic politics of the city.

Infecting Aesthetics: Conflict, Contingency, Contamination

Just as cleanliness unevenly interlinks with the counterpart of contamination, we follow Vittorio Parisi's (2019) proposed term of "infesthetics" as an imperfect conceptual complement to aesthetics. Infesthetics, in a nutshell, describes the "anti-ornamental force of graffiti writing and weeds" whose "adjunctiveness is parasitic, it exploits the architectural element instead of serving it" (Parisi, 2019: 14). Pushing this spectrum of aesthetics/cleanliness and infesthetics/contamination further, we seek to establish an empirically grounded understanding that cleanliness (and contamination) is conflictual. In this sense, we stress the political implications of cleanliness, or simply, the urban politics of cleanliness (McFarlane and Silver, 2016; Cornea, Véron, and Zimmer, 2017; Wittmer, 2022). These multiple political relations and regulations are necessarily ruptured by modes of critique and differences in meaning, value, place, and the identity-related purchases of art.

Furthermore, art and its aesthetics are imbricated in processes of cultural legitimation, valuation, and devaluation (Young, 2016; Millie, 2016) and in struggles over legality (Vanderveen and van Eijk, 2016). This situates art within an aesthetics of authority (Ferrell, 1993), which produces tensions between different political, civic, and judicial stakeholders at the cost of disadvantaging independent and non-formalised positions. These aesthetic tensions bring up questions of *whose* city, and *whose* aesthetics govern the city. With a class- and inequality-attuned perspective on urban civil society movements in India, particularly Mumbai, Yoko Taguchi (2013) assumes a close association between civic values and cleanliness in "middle-class activism". Cleanliness, in her view, can interrelate with forms of colonial, paternalistic leadership over parts of civil society that supposedly need to be educated on how "to be proper citizens" (Taguchi, 2013: 89). In that sense, "cleaning up" can be considered as a pedagogical practice of reclaiming public space, but also as imposing orders, disciplining, and excluding bodies within public space (Fischer and Dürr, 2020; Ghertner, 2010). Moreover, Taguchi quotes Leela Fernandes (2006: 144), who ascribes a "new aesthetics of class purity" and cleanliness to a new Indian middle class. These ordered citizens exemplify an "aesthetics based on the middle class desire for the management of urban space based on strict class-based separations" (Fernandes, 2006: 144). Pulling these insights into the Austrian context, we examine below how seemingly middle-class values (and aesthetics) are functionalised as part of neoliberal and right-wing-leaning political imaginaries of the city that shape the specific conflictual aesthetics of cleanliness in Linz.

As a conceptual ally to these aesthetic politics of cleanliness, and clean politics of urban aesthetics, we draw on a conflictual sense of aesthetics (Marchart, 2019). The latter assumes that there "are" no aesthetics without contingent articulations or sedimentations (i.e. repetition, routinisation, and normalisation) of hegemonic assumptions about what is aesthetically pleasing, desirable, or beautiful. As Marchart elaborates (2019: 23), such conflict-attuned aesthetics need to be attended to as:

> conflictual in a double sense: it conflicts with the aesthetics of the spontaneous ideologists of the art field (the aesthetics of simplistic complexity), and it seeks to work out the political implications of conflictual artistic practice. It is, in this double sense, both a conflicting aesthetics and an aesthetics of conflict.

We consider the latter part of the quotation of importance in sketching a conflictual aesthetics of urban space, and a conflictual aesthetics of street art and graffiti in particular. Within these aesthetics, conflicts are not considered a nuisance or problem to be eradicated, but instead, a driving and constitutive dimension in embracing contingency and contamination rather than cleanliness. For the following discussion of the aesthetic politics of cleanliness in Linz, and its aspiration to become the cleanest (post-industrial) city in Austria, we consider Marchart's conceptual proposition insightful to make clear that street art *antagonises* cultural urban politics and policy. Notably, Austria's third-largest city has not been subject to either sociological or cultural political research and publications (exceptions are John et al., 2009; Schad, 2019; Landau-Donnelly et al., 2023). In light of this lack of in-depth sources, the following introductory city profile sheds light on the historical background, which has led to the contemporary aesthetic politics of clean(ing) Linz.

Linz: Desiring Cleanliness in a Post-industrial Era

Linz, with a population of approximately 209,000 people, is located on the river Danube in Austria's central North. The city is governed by a Social Democratic mayor, whilst its Councillor for Culture and Tourism belongs to the Conservative party. It is the capital city of the province of Upper Austria, and has a strong industrial background, being particularly known for the production of steel and chemicals. One of the city's major enterprises is the steel company *voestalpine AG*, formerly *VÖEST-Alpine AG*, which is not only one of the largest local employers but also coined the city's reputation and identification as "steel city". As such, heavy industry has not only had a profound economic impact on the city's prosperity but also played a key role in residents' identities as "hard as steel" or "steel city children". These discursive tropes still "work well with the city's image", as a local cultural policy researcher and artist declared (Interview 18.5.21). The other influential industrial player had been the former chemical company *Chemie Linz AG*, which existed until 1990. Due to its location in Linz, a large industrial area—the so-called *Chemiepark Linz*—was developed in the harbour area where chemical plants clustered. Particularly, in the context of debates, around environmental and air pollution, these companies played major roles in giving the city a rather "dirty" image.

"In Linz, it stinks!" Framing Cleanliness in Urban Re-branding

During the 1980s, due to the increasingly polluted air, and resulting bad smell and smog holding sway over the city, Linz suffered from a serious image problem. For years, Linz was presented as a dirty industrial city in public broadcasting channels, coining the unflattering slogan "In Linz, da stinkt's!" ("In Linz, it stinks!"). Although political measures to improve levels of environmental pollution had already been initialised in the 1970s—leading to the establishment of the municipal Department for Environment Protection in 1983—the apprehension of negative economic side effects inhibited political decision-making (John et al., 2009: 45). Hence, discussions in the local press were regularly dominated by reports on the misconduct of leading industrial plants and the severe negative health effects on Linz's citizens, particularly children (Oberösterreichische Nachrichten, 1984). In reaction to these reports and the insufficient intervention by political stakeholders, as well as the companies themselves, in late Neue Kronen Zeitung, 1985, local students went on strikes and organised demonstrations with several thousand participants (*Neue Kronen Zeitung*, 1985).

The emergence of these environmental protests increased the pressure on the city government to prioritise ecological issues on their agenda. As a result, during the city council elections in Neue Kronen Zeitung, 1985, the slogan "Linz muss die sauberste Industriestadt Österreichs werden!" ("Linz has to become Austria's cleanest industrial city!") was introduced. This new direction, potentially a forebode of the aesthetic regime of cleanliness to be reinforced later, leveraged a long-lasting effect on future policy directives and decisions in urban and planning-related processes. In sum, the civic-led political contestations of Linz's image as a dirty industrial city did not only elicit extensive public investments in environmental protection measures. It also led to pushing the (financial) public support of arts and culture to the fore. Accordingly, while the aspiration of cleanliness has its roots in primarily environmental, anti-pollution policies against "hard" contamination of air quality, soil, water, and so on, we position the politics of cleanliness more broadly in the aesthetic realms of urban branding, placemaking, tourism, cultural policymaking, and public communication strategies. The imperative for cleanliness influenced urban cultural policy as one of the driving factors in re-branding Linz's collective urban identity (John et al., 2009: 46).

In the early 1990s, the city experienced a marketised renaissance, expedited by large-scale urban marketing campaigns under slogans such as "Linz. Eine Stadt lebt auf!" ("Linz. A city lives up!") or "Linz im Anzug" ("Linz in suit", Figure 5.1). Specifically pushing cultural and economic themes in city branding, Linz presented itself as a dynamic, successful, and future-oriented enterprise with matching corporate design, media, and poster campaigns (Laister, 2004: 194).

Figure 5.1 "*Linz im Anzug* [Linz in suit]", City Marketing campaign, 1989–1994 (Courtesy: Linz City Archive).

An important marketing focus was to foreground cultural events, media, and art and interlink cultural experiences with economic activities such as shopping and visiting restaurants and bars (John et al., 2009: 49). At the same time, the city underwent a significant economic transformation as the state-owned enterprise *VÖEST-Alpine AG* was re-structured and partly privatised in the mid-1990s (in 2005, its privatisation was completed).

Unsurprisingly, these rising market-oriented tendencies also affected the local production and institutional infrastructure of arts and culture. As Linz's rebranding as a clean, attractive city relied heavily on added value through the expansion of its cultural sector, the city facilitated the foundation and construction of two major cultural institutions, the *Ars Electronica Center* (1996) and the *LENTOS contemporary art museum* (2003). Besides these cultural-political investments, during the 1990s, the application for the *European Capital of Culture* was prioritised as part of a long-term cultural development strategy. In 2009, when the city was awarded as *European Capital of Culture*, marketing campaigns focused on the transformation from an industrial city to a "city of culture". Taking these steps also meant a clear orientation towards neoliberal creative city strategies—namely, linking cultural policy decisions with logics of economic urban development (Miles and Paddison, 2005; Ponzini and Rossi, 2010). As a result, Linz became a member of the *UNESCO Creative Cities Network* as a *UNESCO City of Media Arts* in 2014 (UNESCO Austrian National Commission, n.d.).

Who's cleanest? Situating Linz's Local Graffitiscape in Cross-city Competition

Not least because the German word *Provinz* [province] rhymes perfectly with the name *Linz*, it was a common side blow to dismiss Linz as provincial. In our discussion of the aesthetic politics of cleanliness and the conflictual aesthetics of street art, this particular framing of the city is relevant for two reasons: first, because it has had a major impact on handling urban tourism. In light of cross-city competitiveness as an attractive cultural tourist destination, Linz wrestles with cities such as Vienna and Graz. Linz's year as *European Capital of Culture* in 2009 aimed to enhance cultural tourism, marking a turning point according to a social scientist and artist, who has been engaged in researching and exhibiting the evolution of the city's cultural identity:

> Before 2009, Linz was really in a sorry state, in a state of decline. 2009 has brought many new connections (…) Linz Tourism, that is fair to say, still has a hard time marketing Linz as a city of culture. Especially in comparison to Salzburg, Vienna or Graz. But in my opinion, they've done a very good job of moving towards the new, the contemporary, the experimental.
> (Interview 18.5.21)

Over the past few years, Linz's cultural tourism agenda has further become heavily entangled with the city's approach to street art and graffiti. According to a cultural and urban art researcher, who is based in Linz and focuses on graffiti, among other things, urban tourism is "almost aggressively promoted via urban art" (Interview 27.5.21).

Second, the small size and supposedly provincial character have affected the profile and development of Linz's street art and graffiti scenes. Interviewees who live in Linz and work there as street artists describe the scene as quite small (Interview 10.6.21). According to one local graffiti artist, there are only a few writers located in Linz who are doing illegal pieces and just a handful of artists who work with graffiti on legal walls or create commissioned murals (Interview 8.9.21). The relationship between the graffiti scene and the relatively more established (independent) art scene—focused on media and digital arts—is described by an artist and cultural political researcher as contentious, "somewhere between critical and not really interested" (Interview 18.5.21). Nevertheless, in Linz, there are many contact points between graffiti and hip hop. Speaking from his own experience, one interviewee explained that in the 1990s, many graffiti writers were also rappers or DJs (Interview 14.9.21). This strong relation between the two cultural scenes has grown "organically", as another interviewee, with a background in graffiti/urban research, suggests (Interview 27.5.21).

While street-bombing and train-writing were never big issues in Linz's local repertoire, in the 2000s, the scene as a whole was fairly dormant, according to a local cultural practitioner and organiser (Interview 14.9.21). However, over the past decade, the graffiti scene has grown again, or at least people from outside the graffiti scene have this impression: "Graffiti has never been as visible as it is now in Linz", a cultural administrator states (Interview 15.9.21). At the same time, the local scene's profile is also shaped by artists' tendency to move to Austria's capital city Vienna and work there, although this does not necessarily mean higher income. A local street artist who works in Austria and internationally depicts the increasingly market-oriented logics of street art (Interview 10.6.21). While street artists generally depend on private commissions, she encountered that the field is dominated by a high number of newcomers who work for low rates. However, the appeal and possibilities of a large(r) city might not be the only reason for the migration towards Vienna. A local cultural organiser and event manager argues that city-enforced anti-graffiti actions, and police interventions add to graffiti writers experiencing Linz as "too hot" a territory to write in (Interview 25.6.21).

Ordering the "Official" Politics of Cleanliness

The aesthetic politics of cleanliness are shaped and enforced by city authorities via various political measures of socio-spatial control. In addition, they shape a formalistic understanding of what "art" or "aesthetics" are. As laid out above, pursuing cleanliness has become a crucial strategy in Linz's efforts for urban rebranding. "Linz is, after all, the most beautiful or not the most beautiful, the cleanest city in Austria. The Social Democrats are very proud of that", a local artist and social researcher argues (Interview 18.5.21). He traces the transition from "'Linz stinks, it's an awful city' to 'Let's all be extraordinarily clean now'", a seemingly logical development "when it comes to the environment and nature, but this logic has just been transposed to the cultural realm as well" (Interview 18.5.21). Acts of cleaning thus turn into dynamics of cleansing, which instantiate aesthetic conformity and insert a corporate marketing logic into public space.

The material effects of the city's regulative approach towards the use and appearance of public space also manifest in its stance on graffiti and street art. "We definitely have a problem with property damage. So where private property is defaced, public property is also damaged. Then you have to persecute that", a leading cultural administrator argues (Interview 15.9.21). In this sense, opinions about street art and graffiti as defacement, destruction, and an assumed lack of respect for monuments in public space as well as public and private property are justified with police measures and legal charges: "The city has its own budget to clean up" (Interview 15.9.21). However, city authorities pursue a normatively charged approach as "cleaning up" in the form of erasure does not apply to every case of street art or graffiti. "If it fits in 'nicely' somewhere, then yes [street art can stay]" (Interview 18.5.21), the situation is described by a local artist and social researcher. A local street artist explains that murals are increasingly commissioned by the city itself (Interview 6.10.21), a development that is due in no small part to the increasing popularity and high-profile staging of the street art gallery *Mural Harbor* (see below). However, this interviewee further describes that the Committee for Urban Design, which has many representatives from the far-right-wing party FPÖ, can veto or reject street art proposals on the basis of aesthetic criteria that favour cleanliness. This hints at the way the aesthetic regime of the "clean post-industrial city" has started to manage graffiti and street art over space and place. Thus, whether or not it is assumed clean or unclean, and the types of graffiti that are welcome, is governed by the sense of place that city authorities attempt to create. Members of the FPÖ, especially expect the (inner) city to "remain clean at all times", a cultural manager involved in the organisational team of *Mural Harbor* states (Interview 25.6.21). But how the status of "clean space" is being interpreted—and executed—is based on contingent and ideologically charged judgements. These judgements might be slow to change over time, assigning different aesthetic priority values to the various forms of street art and graffiti.

Across the spectrum of political parties but also in civil society, graffiti tags are a polarising topic, and often considered a nuisance and cause for annoyance (Kreslehner, 2020: 13). Not only police forces but also private individuals act as self-authorised cleaners of illegal graffiti and stickers, removing them self-directedly or reporting it under the category "graffiti" to the city-owned website www.schauauf.linz.at (Kreslehner, 2020: 13). A member of a cultural initiative that has organised graffiti-related events in Linz considers city authorities to be not very reflexive and having little idea about the interconnecting occurrence of commissioned public art and unsolicited graffiti. This interviewee imitates the city's logic: "'Murals yes, please! And make nice images!' But you are not going to get colourful cute kittens without taggers responding to it" (Interview 11.10.21). This quotation again illustrates the contingency and conflictuality that pervades the hegemonic politics of art and aesthetics.

"Very pleasing, but zero message": Tracing Graffiti in Linz

Notably, the City of Linz does not pursue a zero-tolerance policy against graffiti as it can be found in municipalities in Norway, Sweden, Australia, and the United States (Young, 2010; Kimvall, 2015; Shobe and Conklin, 2018; Arnold, 2019). There are no aggressive anti-graffiti campaigns. Still, the city's practices are regulative, exerting bureaucratic control over location and access to spaces, hence enforcing one specific conception of aesthetic politics of street art that is allowed and wanted. A street artist based in Linz recounts that public authorities, funding bodies, planners, and policymakers prefer the term "street art" over "graffiti"; thus, artists try to avoid the latter term (Interview 10.6.21). This glitch of language not only points to the classificatory ambiguities one necessarily encounters when distinguishing and defining street art and graffiti; a task quite difficult due to the variety of meanings, materialities, and forms, as well as legal, cultural, and social implications of (licit and illicit) urban mark-making (Schacter, 2017: 104). It also illustrates that the categorisation into "street art" (assumed to be legal) and "graffiti" (assumed to be illegal) is essential for these practices' "process of cultural legitimation" (Young, 2017: 49). What Alison Young frames as "progressive familiarization of street art" (Young, 2017: 49–50) needs to happen in distinction to graffiti and the "distaste" graffiti is often met with. In the end, these value statements lead to making street art legally and culturally permissible—or not (Schacter, 2014: 164). The hegemonic politics and policies of this aesthetic regime thus favour consensual—in other words *clean*—forms of public art that are legitimised by being deemed visually appealing and artistic expressions. They stand in stark contradiction to "ugly", ambivalent, or politically controversial manifestations of graffiti. However, material traces of graffiti in public space would embrace the logic of contamination (Fridrik, 2023). In sum, the local production and reception of graffiti and street art in Linz appear to be situated within a conflictual dynamic between the aesthetic politics and policies of urban branding and an imperative for cleanliness on the one hand, and the more subversive logics of subcultural organisation and articulations of street art on the other.

Inner-city murals in Linz are mostly curated co-operations with the initiative *Mural Harbor*, a key stakeholder within the local street art scene. They are mainly created by internationally renowned artists "with more or less of a graffiti background (…) very pleasing, but zero message" (Interview 27.5.21), as a local urban art researcher argues. This statement suggests that the large-scale murals located in the city centre are rather representing a normative and decorative design than radical artistic ideas. Although the city offers legal walls on which writing and painting are permitted, they are not present in central locations of the city (Spraycity, n.d.) nor are they deemed attractive for the artists they are designed for. "Legal walls in Linz are bad", several graffiti stakeholders say in interviews. On the one hand, "the walls are poorly located, e.g., in pedestrian underpasses where there is not enough light to work or take photographs" (Interview 14.9.21), a feature which has become crucial when doing graffiti and street art. Furthermore, it can also be argued that legal walls clearly confine (or tame) graffiti writing to designated spaces, imposed by official authorities (see McAuliffe, 2010). On the other hand, according to a local graffiti writer, legally provided space is too limited, resulting in "disputes within the scene" (Interview 9.8.21). As not enough space is provided to write concurrently, practices of palimpsest-like layerings of different pieces lead to a rather quick overwriting of also technically more elaborate works which would usually remain "untouched" for a certain while. Another street artist indicates that the area around the main train station would offer adequate space in the form of empty buildings and walls, but those have been sealed off and guarded by the city for more than 10 years. These are areas where "nothing [in terms of graffiti art] may and will actually take place in the near future", as this artist criticises (Interview 10.6.21).

However, over the past decade, the city has become generally more tolerant towards graffiti and street art practices. Authorities are providing additional legal walls and actively supporting corresponding projects—even in the city centre—with public funds; as long as they are moving within a legal framework, an interviewee working in cultural administration clarifies (Interview 15.9.21). An interesting compromise between graffiti writers and the city has been established in the form of the largest *Hall of Fame* in Linz, located at a central area next to the river Danube. A well-established musician and cultural organiser ascertains that "although writing there is only legal with a city-issued notice, accounts of graffiti are more tolerated there than in other central locations" (Interview 14.9.21). And yet, as another interviewee with a background in cultural event organisation points out, the question of whether this is a sign of "trust and the desire to interfere or of the beautification of the grey wall" remains unanswered (Interview 11.10.21).

In light of the ambivalent valuations of street art, a cultural administrator in Linz argues:

> We're caught in a conflict again. That art has to be well-behaved, it has to be beautiful, that's the classic art discussion (...) is the ugly even allowed to be exhibited at all, or is that unsightly or repulsive? Of course, everything should be allowed, but even advocates of street art reach their limits when their own interests are negatively affected by street art.
>
> (Interview 15.9.21)

This statement from a leading local cultural administrator reveals the partial openness and receptiveness to street art ("everything should be allowed"). Yet it also implicitly reigns in the underlying middle-class values ("their own interests") that might stretch to aspects such as the uninhibited intactness of private property and real estate. The criterion of permissibility relates to the city-led discourse on what is allowed versus what is possible in the urban streetartscape (Landau-Donnelly, Forthcoming). Put differently, clean(er) (i.e. less ambivalent or less contaminated) street art forms are appreciated, dismissing other forms such as tags as "smears". Similarly, a Linz-based graffiti writer and cultural organiser argues that only certain types of street art are welcomed in public space: "When the graffiti is pretty, it's fine, but the smears of graffiti are not welcomed; they are automatically considered vandalism" (Interview 14.9.21). While not all graffiti have an explicit (or at least legible) political message, the act of graffiti-writing itself may or may not be considered political (Waldner and Dobratz, 2013). In that sense, graffiti can point to, and thus destabilise, the cleanliness of urban orders and public space—in other words, it can (aesthetically) *infest* them. Historically, graffiti might be seen as inherently contesting the notion of property (and maybe value, too), precisely by appropriating/disturbing/devaluing it, but more recent graffiti art such as Banksy's sellable pieces might suggest that also graffiti art can become commodified.

Hotspot Harbour: Displacing Street Art to Urban Margins

Opened in 2012, and in cooperation with the municipal holding company for public infrastructure Linz AG, the private initiative *Mural Harbor* invited national and international street artists to create large-scale murals in the harbour area. Initially, the project was privately funded and only received public cultural funding—by the City of Linz as well as the Province of Upper Austria—by the end of 2022 (Gruber, 2023). However, *Mural Harbor*'s impact on the city's cultural-tourist self-conception and branding can be retraced earlier through cultural political discussions held during local council meetings, where it was mentioned as the "City's new trademark" (Linzer Gemeinderat, 2019: 443) by the City Councillor for Culture. As owner of the physical space, the Linz AG harbour administration provided necessary street art walls and spaces, turning the harbour area into a hotspot for street art and graffiti in Linz; with several interview partners labelling the area a "tourist factor". According to a local cultural administrator, the city's attitude towards graffiti and street art has changed over the past years, from a "criminal countersteering against an illegally operating scene" to actively providing more legal spaces for graffiti writers (Interview 15.9.21). An interviewee, who is active in the graffiti scene himself, links higher levels of tolerance and more cultural political attention to the media impact of *Mural Harbour* for both Austrian and international tourists (Interview 9.8.21). For illegal sprayers, however, *Mural Harbor* is "the last place which would be interesting to write in", which *Mural Harbour* organisers are well aware of (Interview 25.6.21).

Nevertheless, members of the local graffiti/street art scene express criticism of how the project is run and developing over time. On the one hand, according to a Linz-based graffiti writer, "very few local artists are shown on site" (Interview 8.9.21). Linz's tourism approach is considered to advertise urban art "almost offensively, aggressively" while local artists "do not get a piece of the cake", an urban and cultural researcher states (Interview 27.5.21). There was almost a mutually parasitic relation between *Mural Harbour* and the City of Linz, in which the latter advertises (for) itself by providing investment support to *Mural Harbour*. Now, in light of the recent first-time public funding for *Mural Harbour* (see Gruber, 2023), this complex conflictual relationship might change, once again. On the other hand, the use of available walls has apparently come to a gridlock, a local street artist criticises:

> Space was used here, but now, nothing happens anymore. All surfaces are covered, they are no longer painted anew [...] It is a cow that is being milked, because it now runs effortlessly, but nothing new arises.
>
> (Interview 10.6.21)

According to a local social researcher and artist, the harbour area's "interstitial places and temporary brownfields" could be used more proactively to experiment with place-making and art in the urban realm (Interview 18.5.21). However, this interviewee is not very hopeful that Linz's harbour will become "the next big cultural development zone". Large-scale urban development plans for the harbour area in Linz affect, among other locations, the space *Mural Harbour* has occupied. A massive ongoing reconstruction project named "Projekt Neuland [Project New Ground]", will significantly alter the rough, not-so-clean feel and aesthetics of the post-industrial harbour area, as it will demolish buildings, including the complex of *Mural Harbour*. This area will be replaced by new office spaces, hotels, public green spaces, and some residential areas. According to the website of the holding company Linz AG—which owns the entire commercial harbour area—*Masterplan Hafen* Linz (n.d.) aspires to interlink a "cultural axis" with an "industrial axis" within this newly built quarter.

While space for culture is planned, "Projekt Neuland" also inheres a conflictual dimension as it will sooner or later displace the graffiti art pieces, which are already in place via upscaling and gentrification (Schacter, 2014). As a graffiti writer from Linz summarises: "The artists can vent in the 'rotten' harbour, because sooner or later those buildings will be demolished anyway" (Interview 8.9.21). While *Mural Harbour* has upscaled the area as well, it is also perceived as a peripheral location. A Linz-based cultural and urban researcher observes:

> I always had the impression that policymakers think 'It's okay when the kids can make a little dirt out there and play, but please not in the city centre' (...) the guys from *Mural Harbour* probably see this differently, because they have developed the project out of their own initiative, but I felt like it was there quasi-allowed also as a gentrification measure for the development of the harbour area.
>
> (Interview 27.5.21)

Again, the aesthetic politics of cleanliness indicate, or decide, which art is in and out of place, both literally and figuratively. In this specific instance, the spatial politics of the street seem to outweigh the question of whether street art is deemed to be art or not. *Mural Harbour* offers a contaminated aesthetic politics of space, and simultaneously, a contaminated space of aesthetic politics—thus, a space where the logics of official and self-articulating street art stakeholders meet and clash. More precisely, the contamination *takes place* in the non-central space that *Mural Harbour* claims, and thus intersects with or infests the aesthetic politics that defer the production of street art into that specific area. Notably, the space of contaminated aesthetic politics manifests in both sanctioned and unsolicited street art in the area. In sum, contamination consists of the conscious, yet conflictual encounter between different stakeholders, opinions, or valuations of art whereas, in contrast, the logic and politics of cleanliness would refuse, negate, or eliminate this polyphony, ordering street art into urban spaces designed for upscaling and tourist consumption.

A(nta)gonising Cleanliness: "Linz ist Linz" *Promotion Video*

A more recent account of the city's tourist branding strategy was released by the *Visit Linz* tourism agency in August 2021 on the agency's YouTube channel (Visit Linz, 2021). Their promotion video "Linz ist Linz" ("Linz is Linz"), which registered over 600,000 views after one year online, self-ironically and bluntly talks about certain trends, local stereotypes, and prejudices—for example, a rather rash tone, snorting cocaine, and hostility towards foreigners. The image campaign led to polarising public debates as, on the one hand, it caused critical and even dismissive reactions from city officials—including the mayor—who accused the video of insulting the City of Linz (Winkler-Hermaden, 2021). On the other hand, the campaign was rather successful in creating public attention and was even awarded at the *Cannes Corporate Media & TV Awards 2021* (Der Standard, 2021). Geared towards turning Linz into a "hidden" tourist attraction, or destination off the beaten path, a leading local tourist manager states in an interview that Linz is "not Disney (…) but a tourist-appropriate city without knick-knacks or clichés" (Interview 5.11.21). According to this interviewee, the video creates "a moment of identification with the city for a lot of people (…) it was about capturing authenticity, to not apologise what Linz is (…) the video was made for open-minded people and those people also received it well" (Der Standard, 2021). He argues that "authenticity has a bright future ahead", which needs to be maintained with care regarding the value, or valorisation of authenticity (Der Standard, 2021). By means of creating a consumable product, "edgy" tours and excursions around architecture and street art have multiplied in the last years, the interviewee indicates (Der Standard, 2021).

Besides camera shots on murals which are part of *Mural Harbour*, the "*Linz ist Linz*" promotion video also prominently shows Linz-related graffiti tags like "*Linz stinkt*" ("Linz stinks") and "Linz is lost". Together with street art, the infesthetics of tagging is thus included (and marketed) as a representative feature of the city. "Our city is everything but perfect" the speaker says, accompanying a tracking shot across tagged buildings and a large "dirty" construction site (Visit Linz, 2021, 2:15). From a conflict-oriented perspective, this video de-antagonises some of the negative stereotypes of the city and tries to re-appropriate or "own them" them by making fun of itself, but also by including subcultural voices that are not usually portrayed in glossy marketing campaigns. The video, as well as the debates around it, hint towards the contingency of aesthetic orders and politics of cleanliness/contamination, particularly when being entangled in capitalist logics of marketisation. According to a local researcher and artist, "cleanliness is always good, a clean industrial city, a clean cultural city, but that is all a bit overambitious, everything has to be particularly tidy" (Interview 18.5.21). Street art forms are considered problematic when they "play with dirtiness" (Interview 18.5.21). Another interviewee—a graffiti writer—says "No thanks to a clean city!" and argues that "a city is only alive when it is colourful" (Interview 9.8.21). In this view, while colourfulness may not be the neat counterpart to cleanliness, a city does not need to be perfect or clinically clean. With regards to the latter, a Linz-based cultural researcher problematises the ideological, partially Western-centric underpinnings of cleanliness, stating that "a racist colonialist hygiene discourse is brought in relation with graffiti" which is not culturally universal (Interview 27.5.21). In other words, graffiti—just as cleanliness—can be considered aesthetically problematic in certain locations while appreciated in others in other contexts.

Synthesis and Conclusion: Conflictual Aesthetics of Cleanliness

Within this chapter, we have sketched the political implications of urban aesthetics focused on cleanliness and contamination, interspersed with conflicts. In doing so, we have offered contamination as a conceptual counterpart to the notion of cleanliness. In this discontinuous spectrum, both notions are constantly contaminated with each other's hegemonic or counter-hegemonic significations and assumptions about art. With the help of infesthetics and conflictual aesthetics, we have advanced an understanding of the inherently conflictual underpinnings of urban policies that plan or interfere with the production, appearance, and (dis)placement of street art. But the "proper" implementation of the *aesthetic politics* of the city is an impossible task as there are always different readings and practices of cleaning, cleansing, and contamination at play.

The conflictual constellations in Linz are materialising between city-led efforts at culture-led urban regeneration and place-branding, touristic valorisation, and marketisation of street art culture, on the one hand, and self-organised and often unsolicited responses and commentaries on these appropriations on the other. The sought-after hegemony of cleanliness is challenged by (counter-)hegemonic articulations of contamination while, again, both imaginaries of contamination and cleanliness are contested and dependent on the respective positionality within the current hegemony. Put bluntly, one's vision of cleanliness can be another's nightmare of clinically sterile urbanism, and vice versa; one's dream of graffitied urban surfaces becomes another's anxiety-producing scenario of the urban fabric.

By considering the neoliberal smoothening, or even "cleansing" of urban public space (Rose, 2017), we have foregrounded not only the politically charged, but exclusive implications of the "clean" rationale of city governments executed via local culture-led development, tourism, planning, and street art initiatives. In this sense, it can be argued that cleanliness can lead to the sanitisation, exclusion, and criminalisation of urban practices that do *not* align with the regime of cleanliness, order, and maintenance (Denis and Pontille, 2020; Pow, 2009; Wall 2022). To broaden and substantiate these observations, further research has to be done on how urban planning and tourism agencies problematically incorporate policy idea(l)s such as order, civility, or safety into managerial strategies that culminate in an aesthetic politics of cleanliness. In this regard, it needs to be examined how hegemonic aesthetics of cleanliness shape and reinforce an urban politics of exclusion, at the cost of marginalising vulnerable populations such as people who use drugs or alcohol, those who are asking for money in public space, or those writing into public space unsolicitedly.

However, such a regime of cleanliness, often wedded to safety and middle-class values, cannot ever be fully clean. Briefly, while logics of cleanliness and clean(s)ing often seek to de-antagonise problems and inequities in the urban realm, an infesthetics of contamination bears the potential to unsettle aesthetic politics of cleanliness. Such conflictual aesthetics of cleanliness can critique practices of cleansing and controlling, which are assembled under the empty signifier of "cleanliness". Of course, calls for cleanliness can also be conducive to alleviating urgent environmental crises such as air pollution, smog, and non-degradable waste in urban contexts facing overpopulation, traffic congestion, the spread of diseases, and so on. While our empirical grounds for theorising the conflictual aesthetics of cleanliness derive from a Western context, we are curious to see further engagements with this conceptual trope in other contexts—until then, we are cautious of the specific geopolitics of cleanliness that may play out very differently in non-Western mega cities and metropoles across the globe.

Ultimately, when politics of cleanliness are ideologically wedded to socio-spatial exclusion, cleansing, and neoliberal upscaling, they can contribute to de-politicising the urban fabric by taming diversity and encounters with difference (Valentine, 2008). In sum, we argue for a more "contaminated" and political understanding of street art. With this, we signal the need for more theory-building and activist attention to conflict, politics, and aesthetics in cities which are never, and will never be, clean to begin with—but are always already infested with the never-ending varieties of expressions of "the political".

References

Arnold, E. (2019). 'Aesthetics of zero tolerance'. *City: Analysis of Urban Trends, Culture Theory, Politics, Action*, 23 (2), pp. 143–169. https://doi.org/10.1080/13604813.2019.1615758

Avramidis, K. and Tsilimpounidi, M. (Eds.), (2016). *Graffiti and Street Art. Reading, Writing and Representing the City*. London/New York: Routledge.

Bengtsen, P. and Arvidsson, M. (2014). 'Spatial justice and street art'. *NAVEIÑ REET: Nordic Journal of Law and Social Research*, Special Issue on Law and Art, 1 (5), pp. 117–130.

Blanché, U. (2015). 'Street art and related terms: Discussions and working definition'. *Street Art and Urban Creativity*, 1 (1), pp. 32–39.

Bruce, C. F. (2017). 'Public art, affect, and radical negativity: The wall of daydreaming and man's inhumanity to man'. *Subjectivity*, 10, pp. 223–241. https://doi.org/10.1057/s41286-017-0023-0

Cartiere, C. and Zebracki, M. (2016). *The Everyday Practice of Public Art. Art, Space, and Social Inclusion*. New York/London: Routledge.

Chalana, M and Hou, J, (Eds). (2016). *Messy Urbanism. Understanding the "Other" Cities of Asia*. Hongkong: Hongkong University Press.

Cornea, N., Véron, R. and Zimmer, A. (2017). 'Clean city politics: An urban political ecology of solid waste in West Bengal, India'. *Environment and Planning A: Economy and Space*, 49 (4), pp. 728–744. https://doi.org/10.1177/0308518X16682028

Denis, J. and Pontille, D. (2020). 'Maintenance epistemology and public order: Removing graffiti in Paris'. *Social Studies of Science*, 51 (2), pp. 233–258. https://doi.org/10.1177/0306312720956720

Der Standard. (2021). 'Ironische Werbekampagne "Linz ist Linz" in Cannes ausgezeichnet'. *Der Standard*, 19 November. https://www.derstandard.at/story/2000131282052/skurrile-werbekampagne-linz-ist-linz-in-cannes-ausgezeichnet. Accessed 14 April 2023.

Dikeç, M. (2015). *Space, Politics and Aesthetics*. Edinburgh: Edinburgh University Press.

Douglas, M. (1966). *Purity and Danger: An Analysis of Concepts of Pollution and Taboo*. London: Routledge and Kegan Paul.

Fernandes, L. (2006). *India's New Middle Class. Democratic Politics in an Era of Economic Reform*. Minneapolis: University of Minnesota Press.

Ferrell, J. (1993). *Crimes of Style: Urban Graffiti and the Politics of Criminality*. London: Routledge.

Ferrell, J. (2016). 'Graffiti, Street Art and the Dialectics of the City'. In *Graffiti and Street Art. Reading, Writing and Representing the City*, edited by Konstantinos Avramidis and Myrto Tsilimpounidi. London/New York: Routledge, 27–38.

Fischer, J-M. and Dürr, E. (2020). 'Caring for the "good" city in ethical terms: Graffiti removal in Auckland, Aotearoa/New Zealand'. *New Zealand Sociology*, 35 (1), pp. 49–69.

Fridrik, S. (2023). 'Teaching beyond control? On situating a fence and the agonizing effect of graffiti-based cultural practices as challenge and chance for museum education'. *Art & The Public Sphere*, 11 (1), pp. 69–86.

Gemeinderat, Linzer [Municipal Council Linz]. (2019). 'Protokoll über die 3. Sitzung (Budget) des Gemeinderates der Landeshauptstadt Linz'. *Stadt Linz*, December 16. https://www.linz.at/Politik/GRSitzungen/Content/Wortprotokoll_pdfs/3_Sitzung.16.12.2021.pdf. Accessed 10 April 2023.

Ghertner, A. (2010). Calculating without numbers: Aesthetic governmentality in Delhi's slums, *Economy and Society*, 39 (2), pp. 185–217.

Ghertner, A. (2015). *Rule By Aesthetics. World-Class City Making in Delhi*. Oxford: Oxford University Press.

Gruber, R. (2023). 'In der Hafen-Galerie werden einige Werke heuer für immer verschwinden'. *Oberösterreichische Nachrichten*, January 17. https://www.nachrichten.at/oberoesterreich/linz/in-der-hafen-galerie-werden-einige-werke-heuer-fuer-immer-verschwinden;art66,3775442. Accessed 10 April 2023.

Hawkins, H. and Straughan, E. (2015). *Geographical Aesthetics: Imagining Space, Staging Encounters*. New York/London: Routledge.

Hoppe, I. (2020). 'Graffiti is back in town: Critical approaches to visual and spatial practices in Berlin'. *Journal of Urban Cultural Studies*, 7 (2–3), pp. 199–223. https://doi.org/10.1386/jucs_00026_1

Iveson, K. (2010). 'Graffiti, street art and the city'. *City: Analysis of Urban Trends, Culture Theory, Politics, Action*, 14 (1–2), pp. 26–32.

John, M. et al. (2009). Stadt im Glück: Forschungstexte. https://liqua.net/liq/linz-stadt-im-glueck/. Accessed 20 September 2022.

Kimvall, J. (2015). *The G-Word: Virtuosity and Violation, Negotiating and Transforming Graffiti*. Arsta: Dokument Press.

Kreslehner, K. (2020). 'Hidden Town – Die Stadt in der Stadt. Oder: Wohin man den Fokus lenkt.' In *GRAFFITI & BANANAS. Die Kunst der Straße*. [Exhibition catalogue]. NORDICO Stadtmuseum Linz, 4 September 2020–21 March 2021.

Laister, J. (2004). *Schöne neue Stadt. Produktion und Rezeption postindustrieller Stadt-Bilder am Beispiel von Linz an der Donau. Österreichische Kulturforschung 4*. Münster: LIT Verlag.

Landau, F., Pohl, L. and Roskamm, N. (2021). *[Un]Grounding: Post-Foundational Geographies. Sozial- und Kulturgeographie*. Bielefeld: transcript.

Landau-Donnelly, F. Forthcoming. Political Difference of public art: Exploring contested murals in Vancouver's China Town.

Landau-Donnelly, F, Schad-Spindler, A. and Fridrik, S. (2023). Conflictual cultural politics: Unpacking local tensions in three Austrian cities. *International Journal of Cultural Policy*, 30 (3), 341–356: 10.1080/10286632.2023.2203723

Linz AG. n.d. *Masterplan Hafen Linz*. https://www.linzag.at/portal/de/ueber_die_linzag/projekte/projekt_neuland/masterplan_hafen_linz. Accessed 10 April 2023.

Luger, J. and Ren, J. (2017). *Art and the City. Worlding the Discussion through a Critical Artscape*. New York/London: Routledge.

McFarlane, C. and Silver, J. (2016). 'The poolitical city: "Seeing Sanitation" and making the urban political in Cape Town'. *Antipode*, 49 (1), pp. 125–148. https://doi.org/10.1111/anti.12264

Marchart, O. (2019). *Conflictual Aesthetics: Artistic Activism and the Public Sphere*. Berlin: Sternberg Press.

Markusen, Ann and Gadwa, Anne. (2010). 'Arts and culture in urban or regional planning: A review and research agenda'. *Journal of Planning Education and Research*, 29 (3), pp. 379–391.

McAuliffe, C. (2010). 'Graffiti or street art? Negotiating the moral geographies of the creative city'. *Journal of Urban Affairs*, 34 (2), pp. 189–206. https://doi.org/10.1111/j.1467-9906.2012.00610.x

McLean, H. (2014). 'Digging into the creative city: A feminist critique'. *Antipode*, 46 (3), pp. 669–690. https://doi.org/10.1111/anti.12078

Miles, S. and Paddison, R. (2005). 'Introduction: The rise and rise of culture-led urban regeneration'. *Urban Studies*, 42 (5–6), pp. 833–839. https://doi.org/10.1080/00420980500107508

Millie, A. (2016). 'Urban interventionism as a challenge to aesthetic order: Towards an aesthetic criminology'. *Crime, Media, Culture*, 13(1), 3–20. https://doi.org/10.1177/1741659016631609

Neue Kronen Zeitung. (1985). 'Tausende Schüler streikten wegen des Smogs'. *Neue Kronen Zeitung*, 7 December, pp. 8–9.

Oberösterreichische Nachrichten. (1984). 'Linz: Luft weiter zum Schneiden. Hustenanfälle plagen Kinder'. *Oberösterreichische Nachrichten*, 20 October, pp. 5.
Parisi, V. (2019). 'Haunting the neoliberal city. Illegal graffiti and the "infesthetics" of urban interstices'. *Lo Squaderno: Explorations in Space and Society*, 54, pp. 11–15.
Ponzini, D. and Rossi, U. (2010). 'Becoming a creative city: The entrepreneurial mayor, network politics and the promise of an urban renaissance'. *Urban Studies*, 47, pp. 1037–1057. https://doi.org/10.1177/0042098009353073
Pow, C-P. (2009). 'Neoliberalism and the aestheticization of new middle-class landscapes'. *Antipode*, 41 (2), pp. 371–390. https://doi.org/10.1111/j.1467-8330.2009.00677.x
Rancière, J. (2010). *Dissensus: On Politics and Aesthetics*. Continuum International Publishers.
Riggle, N.A. (2010). 'Street art: The transformation of the commonplaces'. *The Journal of Aesthetics and Art Criticism*, 68 (3), pp. 243–257.
Rose, J. (2017). 'Cleansing public nature: Landscapes of homelessness, health, and displacement'. *Journal of Political Ecology*, 24, pp. 11–23.
Sacco, P. L., Ferilli, G., Blessi, G. T. and Nuccio, M. (2013). 'Culture as an engine of local development processes: system-wide cultural districts I: Theory'. *Growth and Change* 44(4), pp. 555–570.
Schacter, R. (2014). 'The ugly truth: Street art, graffiti and the creative city'. *Art & The Public Sphere*, 3 (2), pp. 161–176.
Schacter, R. (2017). 'Street Art is a Period, PERIOD: Or, Classificatory Confusion and Intermural Art'. In *Graffiti and Street Art. Reading, Writing and Representing the City*, edited by Konstantinos Avramidis and Myrto Tsilimpounidi. London/New York: Routledge, pp. 103–118
Schad, A. S. (2019). Cultural governance in Österreich. *Eine Interpretative Policy-Analyse Zu Kulturpolitischen Entscheidungsprozessen in Linz Und Graz*. Bielefeld: transcript.
Schad-Spindler, A., Fridrik, S., and Landau-Donnelly, F. (2023). Conflictual consensus in Austrian cultural politics: Urban cultural policy research at the intersection of agonism and situational analysis. *Forum Qualitative Sozialforschung / Forum Qualitative Research*, 24(2). https://doi.org/10.17169/fqs-24.2.4068
Schad-Spindler, A., Landau-Donnelly, F., Fridrik, S., and Marchart, O. (2023). *Konfliktuelle Kulturpolitik*. Springer.
Shobe, H. and Conklin, T. R. (2018). 'Geographies of graffiti abatement: Zero tolerance in Portland, San Francisco, and Seattle'. *The Professional Geographer*, 70 (4), 624–632. https://doi.org/10.1080/00330124.2018.1443476
Spraycity. (n.d.). 'Legal Walls Oberösterreich'. https://spraycity.at/legal-walls-oberoesterreich/. Accessed 12 April 2023.
Taguchi, Y. (2013). 'Civic sense and cleanliness: Pedagogy and aesthetics in middle-class Mumbai activism'. *Contemporary South Asia*, 21 (2), pp. 89–101.
UNESCO Austrian National Commission. (n.d.). 'UNESCO Creative Cities'. https://www.unesco.at/kultur/unesco-creative-cities. Accessed 13 April 2023.
Valentine, G. (2008). Living with difference: Reflections on geographies of encounter. *Progress in Human Geography*, 32 (3), pp. 323–337. https://doi.org/10.1177/0309133308089372
Vanderveen, G. and van Eijk, G. (2016). 'Criminal but beautiful: A study on graffiti and the role of value judgments and context in perceiving disorder'. *European Journal on Criminal Policy and Research*, 22, pp. 107–125.
Visit Linz. (2021). 'Linz ist Linz'. 4 August. Available at: https://www.youtube.com/watch?v=ilVoxX7MDGc. Accessed 14 April 2023.
Waldner, L. K. and Dobratz, B. A. (2013). 'Graffiti as a form of contentious political participation'. *Sociology Compass*, 7 (5) 377–389.

Wall, E. (2022). From Cleaning to Cleansing Maintenance as an Urban Development Practice at Paddington Waterside, London. In: *Unsettled Urban Space Routines, Temporalities and Contestations*, edited by Tihomir Viderman, Sabine Knierbein, Elina Kränzle, Sybille Frank, Nikolai Roskamm and Ed Wall. New York: Routledge, pp. 140–152.

Winkler-Hermaden, R. (2021). 'Polarisierendes Werbevideo: Macht es neugierig auf Linz?'. *Der Standard*, 19 August. https://www.derstandard.at/story/2000129017998/polarisierendes-werbevideo-macht-es-neugierig-auf-linz. Accessed 14 April 2023.

Wittmer, J. (2022). 'Dirty work in the clean city: An embodied urban political ecology of women informal recyclers' work in the "clean city"'. *Environment and Planning E*, 6 (2), 1343–1365. https://doi.org/10.1177/25148486221102374

Young, A. (2010). *The Scene of Violence. Cinema, Crime, Affect*. Abingdon/New York: Routledge-Cavendish.

Young, A. (2016). 'Art or crime or both at the same time? On the ambiguity of images in public space'. In *Graffiti and Street Art. Reading, Writing and Representing the City*, edited by Konstantinos Avramidis and Myrto Tsilimpounidi. London/New York: Routledge, pp. 39–53.

Young, A. (2017). *The Scene of Violence. Cinema, Crime, Affect*. Abingdon/New York: Routledge-Cavendish.

Zieleniec, A. (2017). 'The right to write the city: Lefebvre and graffiti'. https://www.erudit.org/fr/revues/eue/2016-v10-eue03138/1040597ar/. Accessed: 20 September 2022.

6 The Post-Race City

Borders, Freedom, and Imagination

Günter Gassner

Introduction

Berlin and Bristol. 1930s and 2020s. Two halves do not make a whole. But a third part is not needed. Two halves produce more than a whole or, better, they produce an aesthetic that is beyond a limiting thought of wholeness. This is no reinforcement of the phrase: the whole is greater than the sum of its parts. There is no discernible sum. There is also no hierarchy between the two halves, and no harmony exists. The potential of two: a centre cannot be easily created. There is a movement between the two, which can only be perpetuated by means of relations to other forces to which the two halves are connecting. There is no complete whole.

This text is not only about the organismic city, that is imaginaries of the city as an organism. It is, first and foremost, an act of writing *with* as well as *against* such a city. It is an attempt to write the non-organismic city or, better, to write non-organismically to problematise urban injustices. The organismic city is an aesthetic idea and also an aesthetic ideal. If we understand aesthetics as *aisthesis*, as a theory of sensuous perception, and if we emphasise the senses as a form of knowledge production, then one question is how we judge what is shared, common, and valued in and of the city. The organismic city, I suggest, is driven by a desire for harmony, hierarchy, and completeness, and it is based on exploitation, marginalisation, and exclusion. In the organismic city, each place and each "member" is given a static identity, defined in terms of how it performs a function for the city as a whole. The organismic city attempts to maintain its structure in the face of change. It is one that reproduces classist, racist, and sexist structures, behaviours, and thought patterns. I challenge an organismic aesthetic not with the help of factual statements, a chronological account, or a linear argument: these are not approaches that break through an epistemology of power, privilege, and domination (cf. Lennard, 2019). Writing non-organismically is a critical and creative attempt to intervene in violent urban imaginaries with the help of a pendular writing movement that oscillates between two places and two times.

In this chapter, I explore urban aesthetics after race. I discuss relations between borders, freedom, and imagination before examining Nazi imaginaries of a *Reich* after race that relate to an understanding of the city as an organism—what I call the "post-race city". I focus on conceptions in which freedom is not regarded as

DOI: 10.4324/9781003335993-6

the opposite but the result of borders and borderisation and I develop a distinction between an aesthetic of un-bordering and an unbounded aesthetic: while the latter is linked to a fascist creativity that is nothing but annihilation, the former, I propose, relates to a radical imagination that is bounded by racialised lifeworlds.

How does the post-race city relate to current claims about the end of race? Goldberg (2015) argues that today's notion of post-raciality fails to address structural and institutional racisms and their historical legacies. "In removing the terms of racial analysis", he writes, "the postracial undercuts any capacity to account for their conditions of possibility and historical emergence, and also their delimitation" (Goldberg, 2015: 88). The post-racial ends "the possibility of *seeing* racisms and their structures of enablement, their conditions of possibility" (Goldberg, 2015: 88f). To put this differently, the post-racial condition means the *experience* of non-racialism, avoiding questions of structural differentiation. Racial events are individualised and racial outbreaks are seemingly disconnected from each other. Despite important differences between current post-racial claims and the post-race city, in both, racism (structural and institutional, a conception of division that permeates all aspects of life) and everyday experiences of racist events are separated. And in both, raciality is aligned with an attack on the social organism's health.

Based on Notes Made in Berlin, July 2022

Figure 6.1 Picture of a "borders kill" banner fixed to the block of Weserstraße 133–137, taken in May 2022. (Source: Günter Gassner).

Kill Borders (Berlin)

Walking the streets of Berlin. Before I head to an anti-fascist archive in Kreuzberg, I want to explore the neighbourhood when I see a banner that reads: BORDERS KILL: SOLIDARITY EVERYWHERE. Fixed to the block of Weserstraße 133–137, the banner is highly visible from nearby (the building is set back a few metres from the pavement) as well as from afar (it hangs at the top floor balcony of the eight-storey building). Black on white—black capital letters on a white cloth that flaps in the wind—held in check by the weight of two training shoes.

I am in a trendy part of Berlin, with a juxtaposition of marginal gentrifiers, hipsters, working class, left activists, and neo-Nazis. As I move along Weserstraße, from the south-east to the north-west, I keep thinking about relationships between the banner and different borders and different boundaries: political borders, social and cultural boundaries, but also political boundaries and social and cultural borders; metaphorical borders and invisible borders; boundaries as "dead zones" and moreover as zones of death, and borders as intense and intensive (not necessarily extensive) zones of exchange of difference. I think of urban boundaries, borders in cities, and borders between nation-states. Borders as institutions for organising differences.

It is not only the specific urban context—the demography of the neighbourhood, styles and materialities of buildings, social and economic infrastructures, etc.—that makes the banner interesting to me. I walk in Neukölln at a time of war in Ukraine, at a time when many Germans try their uttermost to reassure themselves that they do have a welcoming culture that is of a deeper, more meaningful nature than the seemingly short-lasting one in the 2015 European "migrant crisis". Perhaps they stand a good chance of succeeding. This time, as news reporters are not getting tired of emphasising, it is mostly women and children who had to leave their husbands and dads behind, who need refuge. This time, Germans welcome those who, by and large, want to return home as soon as possible. This time, as I have overheard a comment in a supermarket, it is not young men who will exploit our social services and rape our women and daughters.

The banner's message is stirring at a time when many in Western democracies—in the so-called "free world"—seem to agree that an attack that violates national sovereignty must not be tolerated. This is a message at a time when the quantitative majority of Germans support economic sanctions against the aggressor and humanitarian help for the attacked—whilst arguing that this is not enough. Precisely because of "our history" and the mantra "never again", Ukraine needs to be equipped with strong weapons. Killing borders because borders kill? What does it mean to fight not merely the violence *against* a nation-state but the violence *of* the nation-state? With regard to Ukraine, there is no solidarity everywhere in support of the abolition of borders. On the contrary, there is solidarity with someone because of an attack on their borders, that is there is solidarity based on—reproducing and reinforcing—the virtue of borders.

K-Fetisch (Berlin)

Walking on Weserstraße, Wildenbruchstraße is one of those cross streets that creates an urban boundary, perhaps even a border (cf. Sennett, 2011). The socio-spatial condition of the street changes abruptly. A lively but predominantly residential street turns into a leisure area, full of bars, restaurants, coffee shops, boutiques, bookshops, and toyshops. K-Fetisch is located at the north corner of the intersection of Weserstraße and Wildenbruchstraße. I decide to go inside to have a coffee. Browsing through their website, I learn that K-Fetisch is a collective; a non-hierarchically organised enterprise (café, bar, event space, local communities centre, etc.) that is run by people who "come from different radical left and anti-national contexts" who share a "feminist and anti-racist basic consensus".[1]

Friends point out to me the linguistic joke that is encapsulated in the name. "K-Fetisch" sounds remarkably similar to *Kaffee Tisch* (coffee table). But I like to think that there is another way of understanding the name. One that emphasises the hyphen and asks what the letter "K" refers to and what "fetish" means in the current political situation. Does "K" stand for *Kollektiv* (collective) or for *Kommunismus* (communism)? If fetish is "an object, a non-sexual part of the body, or a particular action which abnormally serves as the stimulus to, or the end in itself of, sexual desire" (Oxford English Dictionary), in what ways are solidarity and non-hierarchical and anti-oppressive ways of living desired and also practised? If fetish means "something irrationally reverenced" (Oxford English Dictionary), what has happened to rationality?

Bristol, August 2022

Despite the hopes held by scholars of race and colonialism in the early twentieth century including W.E.B. Du Bois, racial hierarchy is still with us. One hundred years later, "socio-structurally we remain very much bound by race" (Goldberg, 2015: 159) and this situation exists despite but, in parts, as a result of post-racial claims. Neoliberal post-raciality locates racism at the level of the individual: a system "where race no longer matters and where only individual will and character either enables or hampers a person's life-chances" (Valluvan, 2016: 2242). As Goldberg (2015) emphasises,

> key elements of postracial conception were already implicit in the very comprehension of liberalism and its historical reiterations: the stress on the individual, the moral and political refusal of racial reference, predicated on the denial of group relevance as well as historical disparities in wealth and power.
>
> (27)

As neoliberalism paves the way for autocratic leaders and authoritarian regimes, post-racial individualisation places a renewed emphasis on nationalist discourses.

When will there be an anti-racist city, that is a city that is committed to structural anti-racism and critically un-thinks and un-does racialisation? And how will different and competing conceptions of "freedom" work towards that time? How can a non-racial city be imagined? And, indeed, should it be imagined? There is a danger of imagining a city without race and without racism, as long as there is an outside of—an exterior to—the city; as long as a "free world" of the world defines itself as the opposite of those parts of the world that are not free.

Imperceptible Reason (Berlin)

I am rereading sections from *Necropolitics*, in an attempt to think with and through the BORDERS KILL banner. Borders, Mbembe (2019) alleges, are nowadays being offered as the solution to problems that Western democracies define for themselves: "We must close the borders. Filter those who make it across them. Process them. Choose who we want to remain. Deport the rest. Sign contracts with corrupt elites from the countries of" (Mbembe, 2019: 98f). Borders are "no longer merely a line of demarcation separating distinct sovereign entities" but increasingly "the name used to describe the organized violence that underpins both contemporary capitalism and our world in general" (Mbembe, 2019: 99).

Borderisation is a process in which migrants and refugees are not the main focus of the argument: they are neither being given "proper names nor faces and possess no identity cards" and they "are merely a kind of hollowed-out entity, walking vaults concealed by a multitude of organs, empty yet menacing forms in which we seek to bury the fantasies of an age terrified of itself and of its own excess" (Mbembe, 2019: 101). "Walking vaults" is an uncanny and revealing metaphor with multiple meanings. As vaults, migrants and refugees are being offered as "structures" that hide secrets (the terrorist machine that, like a strongroom, needs to be opened and emptied). As vaults, they are being reified, that is they are being turned into commodities, bearers of economic value (like the treasury that waits to be burglarised). And as vaults, they are objects, bodies that are hardly alive (and soon to be found in catacombs), but not human beings.

Borderisation does not aim to cut Europe off from the rest of the world, but to grant "Europeans alone the privilege of the rights to possession and free movement across the whole of the planet" (Mbembe, 2019: 103). Mbembe appeals to us for a world without borders, but his appeal is a cautious one. In today's "society of enmity", it means "daring to imagine the impossible" (Mbembe, 2019: 104). The relationship between the actual and the virtual (i.e. the relation of possibility and potentiality vis-à-vis the actual), the relationship between what is practised and experienced, on the one hand, and what is (and can be) imagined in relation to it on the other: that remains a key aesthetic concern. Based on a critique of progress (c.f. Benjamin, 2006a), how can we imagine an alternative to a hierarchical, racist, and nationalist world? How can we expand our political imagination? But also, what can be imagined? Referring to Mbembe's call for daring the impossible: which city is unthinkable, but also, what is deemed to be impossible in the city, by whom, and why (see Gassner, 2020)?

> Our phenomenological experiences of the world are being thoroughly shaken up. Reason and perception no longer tally. Panic ensues. We see less and less of what is given to us to see, and more and more of what we desperately want to see, even if what we desperately want to see does not correspond to any given reality. Perhaps more than ever before, others can present themselves to us in a physical and tactile, concrete way, while remaining in ghostly absence in a similarly concrete void, almost as phenomena.
>
> (Mbembe, 2019: 102)

Unbounded imagination (Bristol)

Imagination, for Kant, means "making present what is absent" (Arendt, 1992: 79; see also Arendt, 1978). Imagination can mean that I reproduce something I have seen before, or that I produce something new that I have never seen. However, as Arendt argues, imagination can never be entirely productive and is always linked to the reproductive faculty. This produces two political questions. First, how can I break through an overly narrow imagination of the city, that is how can I break through the constraints of the reproductive faculty? Second, and this question is potentially more dangerous as fascism proves, how can I make sure that the productive faculty does not develop a momentum of its own and becomes "untameable" and completely detached from the reproductive faculty (and especially from racialised lifeworlds)?

The main problem, here, is not that of a limited creativity but of an unbounded creativity. A certain type of creativity beyond or, better, exterior to politics. Hitler, as Gilroy (2000) points out, "emphasises the 'creative' aspects of [a] metapolitical sensibility" (49). Without addressing creativity, fascism as *revolutionary* conservatism cannot be grasped. Does creativity as unbounded imagination come to life again in powerful ways? And is an unbounded imagination not often found precisely in relation to spatially bounded imaginaries and in nationalist discourses in particular? Do we therefore find unbounded imagination of spatially bounded imaginaries as a "companion" of post-racial individualisation?

For both Mbembe and Arendt, it seems to me, imagination is only radical (as in political and structurally transforming) when it remains bounded by experienced realities. Our imagination is unbounded when it disengages with the world, a process of abstraction that closes the world rather than opens it up to creative alternatives. If some lives are turned into phenomena, into objects of the senses rather than encountered as fellow creators, co-inhabitants, and comrades, then this has a violent impact on them and on our imagination, too.

In the first place, the effect is violent not because it limits imagination but because it disconnects imagination from lived experiences. To be sure, this disconnection then has a limiting effect in that it sharply reduces imaginations to similarities, which results in a loss of the ability to engage with radical differences and in the longing for a community of sameness.

Imagination after Auschwitz—after the impossible became possible—is imagination that is bounded in experienced realities precisely in order to constrain violence and destruction (Tyner, 2017). When reason and perception do not tally (Mbembe) and when imagination is not bounded by experienced realities (Arendt), what is the nature of the border that saves imagination from its destructive creative quality while, at the same time, enabling the production of necessary and viable alternatives to the status quo? In the city of enmity, enemies exist because they are being imagined, that is enemies are being created for the city to exist.

Viscera (Berlin)

I am in K-Fetisch again, rereading notes on Mbembe's chapter four with the heading "Viscerality". Searching online, I found out that viscerality might describe the quality of being related to the physical as opposed to the virtual or imaginary world or reality. Yet, is it not precisely relationships between the actual and the virtual or between the possible and the incompossible but relationships between perceptions and imaginations that are at the centre of constructing borders? According to the Oxford English Dictionary, "visceral" refers to something that is related to internal organs ("of the internal organs"). It also refers to an intuitive decision rather than one that is derived by reason. Moreover, "visceralisation" refers to the "spreading of an infection to the viscera; the movement of a pathogen towards the viscera".

Is the city of enmity a body with organs? Is it an assemblage of organs and, at a higher level, an organ in a social (and national) body itself (see Bonta and Protevi, 2006)? Does it consist of organs that are organised in a specific way: centralised, hierarchised, and self-directed? And are those organs under threat of being attacked by an infectious disease? Is it a city that is not driven by a rational response to a potential threat but driven by intuition guided by paranoia? The city of enmity as a body is only a specific selection of what the city can be because it can be organised in various different ways. The city of enmity is only one specific actualisation of virtual bodies. In this specific actualisation, the city is not only a body that assembles organs (while being itself an organ), within it selected human bodies (migrants, refugees, and racialised groups) are being stripped of not only their nature as organisms (i.e. a certain organisation of organs) but also their nature as bodies (i.e. an assemblage of organs). Singled-out bodies are being constructed as dead bodies.

Borderisation, according to Mbembe (2019), is not a war against "singular bodies but rather great swaths of humanity adjudged worthless and superfluous, whose every organ must be specifically incapacitated in a way that affects generations to come" (100). Because it is not a war against singular bodies, the bodies of migration and refugees are constructed as a "hollowed-out entity, walking vaults concealed by a multitude of organs" (Mbembe, 2019: 101). This hollowed-out body that is concealed by organs is a *body with no organs*, but not in Deleuze and Guattari's (2016) sense of a "body without organs". Deleuze and Guattari describe a body that is not organised in a specific way, that is a body without habitual connections that are centralised and hierarchical: a "non-organismic body" (see Bonta and Protevi, 2006: 121). The city of enmity, on the other hand, is a city (body) with bodies with no organs, that is a city where bodies' organs have been forced to stop functioning. These selected bodies do not have organs that are released from habitual patterns but rather these bodies are stripped of the opportunity to develop habitual patterns, that is they are not being given the opportunity to live.

Anti-liberal Freedom (Bristol)

From the city of enmity as a body to the organic order of life in classical fascism. In an essay written in 1934, Herbert Marcuse (2009) shows not only how fascism "struggles" against liberalism and how liberalism operates in and through capitalism (as liberalism sees the free and private initiative of the entrepreneur as the surest guarantee of economic and social progress) but also how economic liberalism creates the authoritarian state, which, in turn, annihilates individual existence. Unlike seeing freedom (including freedom of movement) and security (here as a social, political, and legal construction of borderisation) as being one of the "major contradictions of liberal order" (Mbembe, 2019: 103), borderisation is also linked to freedom, to the liberal order, and to an anti-liberalist concept of freedom and in this sense to an anti-Semitic and racist ordering of the world that refers, as I propose, to imaginaries of the post-race city.

At the core of the entanglement between economic liberalism and authoritarian anti-liberalism are ideas about an organic order of life, which Marcuse explores in relation to fascism's universalism, naturalism, and existentialism. Here we have a society where "the whole is primarily given in its organic segmentation: the numbers serve the whole, which is superordinate to them, but they serve it according to the unique character that appertains to them as members" (Krieck in Marcuse, 2009: 13). A society with a wholeness

> presents not the unification achieved by the domination of *one* class within the framework of class society, but rather a unity that combines *all* classes, that is supposed to overcome the reality of class struggle and thus of classes themselves.
>
> (Marcuse, 2009: 14)

A classless society is being imagined, but one "on the basis of and within the framework of – the existing class society" (Marcuse, 2009: 14). This is a society that is "united by characteristics of 'race', the preservation of whose purity is the condition of the folk's 'health'" (Marcuse, 2009: 15).

Within liberal existentialism "the existential appears essentially as a contrast to the 'normative', i.e. as something that cannot be placed under any norm lying outside it" (Marcuse, 2009: 21). This means that political conditions "are sanctioned here as existential, and within the political dimension it is the relation to the enemy, or war, that counts as simply and absolutely existential" (Marcuse, 2009: 22). When political relationships are regarded as existential ones, when behind all social, economic, cultural, and religious dimensions of life stands total politicisation, then this implies the end of private and public life. The total state is the "only source and object of every individual's duties. Being in this state of duty abolishes the private character of individual existence" (Forsthoff in Marcuse, 2009: 27). In short, "politicizing the concept of existence", Marcuse (2009: 27) alleges, leads to an annihilation of individual existence, and this, in a nutshell, is a concept of freedom that is anti-liberal.

Bodily Freedom (Bristol)

Freedom and authoritarianism are not opposites, but authoritarianism produces a new, different kind of freedom, which liberalism with its individualism cannot stand up to. This anti-liberalist freedom is produced in the form of an "organic totality against individualist dissolution" (Krieck in Marcuse, 2009: 1). The figure and life of the body are crucial here. Freedom not only in relation to economic liberalism and political freedom but also freedom in relation to (and despite) the constraints of the human body. As Gilroy emphasises:

> fascist movements gave the body and its pleasure back to people who had been schooled in the cultural and spiritual assumptions of Christianity and liberalisms. These doctrines told them that their sense of embodiment and corporeal constraint was a stage to be passed through en route to a higher and more valuable sense of freedom associated with the ideas of the soul and the spirit [...] Hitlerism finds and founds a new definition of freedom from an acceptance of being constrained by the body.
>
> (Gilroy, 2000: 175; see Levinas, 1990)

The new definition of freedom relates to a bodily freedom: not the freedom of movement (of bodies), not the freedom of bodies (bodies without organs), but the freedom that results from working through the constraints of the body. While both fascism and Marxism are materialist, it is the former that finds not only constraints but also freedom in the material constraints. "To become conscious of one's social situation is, even for Marx, to free oneself of the fatalism entailed by the situation", Levinas (1990: 67) writes: "The body is an obstacle. It breaks the free flight of the spirit and drags it back down to earthly conditions, and yet, like an obstacle, it is to be overcome". In fascism, freedom is linked to feeling one's body. It involves "becoming aware of the ineluctable original chain that is unique to our bodies and above all accepting this chaining" (Levinas, 1990: 69).

Politicisation and Aestheticisation (Bristol)

In fascism, the body is seen as the key to freedom and not as its hindrance: the human body, the racialised body, the urban body, and of course the national body. Marcuse's total politicisation thesis does not stand in opposition to Benjamin's (2006b) famous description of fascism as the "aestheticizing of politics" (270)—and the ways in which "fascism aestheticizes war" (Neocleous, 1997: 17). Benjamin shows that fascism replaces desires for socio-economic change with a depoliticised spectacular performance of creativity that is annihilation (Gassner, 2021). "Only physical destruction and death on a massive scale become spectacles that are intense enough to satisfy the political craving for transformation without [...] changing the capitalist class structure" (Gassner, 2022: 803). Hence, both Marcuse and Benjamin show the inherently capitalist structure of fascism despite (but also, with the help of) anti-capitalist statements and even resentments. While Marcuse shows that everything becomes politicised but only in the sense that everything becomes seen and imaged as an existential condition, a condition of either life or death, Benjamin shows how the aesthetics of life or death replace a politics of difference. In the end, both aestheticisation of politics and total politicisation mean the end to politics that is not annihilation.

The spectacle of death and destruction on a massive scale operates within the theatre of bodily freedom, that is freedom through rather than despite bodily constraints and chains. There is no need to worry or complain about the sacrifices that one has to make in a time of war. Instead, the time of war can only be celebrated because it is the time of intensified bodily suffering; it is also the time of elevated freedom. According to Gilroy (2000), Benjamin associated the politicisation of art with the "liberatory power of communism" (151). Have we indeed, as Gilroy suggests, been "deprived of that utopian alternative by our understanding of communists' affiliation to the principle of government by terror as much as by knowledge of their parallel enthusiasm for similar heroic, monumental technologies and aesthetic codes" (Gilroy, 2000: 151)? Today, "the race for separation and de-linking" in a world where borders are "no longer sites to be crossed but lines that separate" (Mbembe, 2019: 2, 3), in such a world, is there a freedom that can be associated with the imagination of an alternative world? If an enemy is constructed in order to create a sense of being "more free", what happens when the enemy has finally been destroyed? What is the role of imagining a city after the enemy?

The Concentric City (Berlin)

The concentric model of a city is centralised, hierarchised, and self-directed, and hence, bounded. This is a city with a central point and distinct areas whose positions within the hierarchical urban structure are defined by their relative distances from the central point. At the same time, in the concentric city equality or, better, claims of equality are morphed into an authoritarian figure of power. The concentric city works like a grid in authoritarian space-time: it works through centralised equality, which is why it is hierarchised. And it is bordered; it defines itself through an exterior. It is compact: a compact "us" (urban organic community) that is organised around "him" (*Führer*) and forearmed against "them" (exterior enemies). Spatial translations into urban form can vary. The concentric city can involve a variety of different spatial characteristics, also, but in no way solely, radial lines of movement and ring-shaped public and private, open and closed spaces. This is a model that is authoritarian and liberal/anti-liberal—but not illiberal or, better, liberatory (see antifa's illiberal tactics in opposition to fascism's anti-liberalism). "Solidarity everywhere" means that one has to spiral out of this organisation. Spiralling out of order, as I have suggested elsewhere, can lead to solidarity and liberation; but it can also lead to complete annihilation (Gassner, 2021, 2022). In the first instance, however, the concentric city is the city which fascism spirals into.

Diagram of Imagination (Bristol)

Can we draw a diagram of a wished-for city? For the city of enmity this means, can we diagram the post-race city? A diagram, Deleuze (2015) comments in his work on Foucault, "is the map of relations between forces, a map of destiny, or intensity, which proceeds by primary non-localizable relations and at every moment passes through every point" (32). While it is a relational power without a clear design, form, or function, it is capable of organising elements. The diagram of the post-race city territorialises an urban assemblage in a certain way, or it can reframe an assemblage in a certain way (cf. Deleuze and Guattari, 2016: 163).

When does fascism exist? If fascism is a racist, nationalist, and ultra-nationalist war of annihilation, is there an end to this war? Is the war supposed to end? And if so, is fascism the war or the world after the war? Hitler was confident, at least in his speeches, that "[t]he war will be over one day. I shall then consider that my life's final task will be to solve the religious problem" (Hitler in Burleigh, 2000: 717). This statement not only indicates the Nazis' contradictory relationship to Christianity with regard to freedom but it also suggests that they might have seen war really as the means to an end (the expansion of the German *Lebensraum*, the Jewish problem solved). Yet, classical fascism cannot be understood in this way alone. Fascism's aestheticisation of politics, violence, war, as well as politics as the "right to exist" (which implies the right to kill) points to war as an end in itself; to a gratification of desires to kill, to oppress, and to be oppressed (see Foucault, 2009).

In addition to the freedom gained from the city being in a state of war, there are, nevertheless, imaginaries of the post-war city. These imaginaries of the post-war city are not post-fascist but truly fascist. They are not, against what some fascists might have argued, concrete urban models that wait to be realised. In no way are these justifications of violent means because fascists do not need a justification for that nor a seal of approval. They work through desire production and not solely with the help of the definition of an end goal. "Fascism's persistent presence as both a minor political option and a major cultural reference point is also a reminder that it must be considered in its positivity", Gilroy (2000: 49) suggests. This positivity can be seen in the capacity to creatively produce diagrams of the wished-for city. When I refer to the diagram of the post-race city, then it is the desired production that results from a disconnection between the actual city and the wished-for city that I have in mind.

Diagram of the Post-race City (Bristol)

The Nazis' ambitions in urban planning were threefold (c.f. Gassner, 2023). First, they developed monumental rebuilding plans for Berlin and a few other cities. In an attempt to transform Berlin into "Germania", the comprehensive rebuilding scheme involved the demolition of much of the historical core and the construction of grand public buildings, wide boulevards, and nationalistic monuments. The "Avenue of Splendours", a pedestrianised parade route with vehicle routes underneath, was planned to connect the Great Hall (a monumental, domed building of excessive scale) with a Triumphal Arch, ending in a new train station (Gassner, 2022). Second, Nazis destroyed selected neighbourhoods. The term "redevelopment" was replaced by that of "recovery" [*Gesundung*] (Münk, 1993), that is a process of making the city healthy again through demolition of buildings and the replacement and annihilation of Jews. Neighbourhoods in which poor urban conditions did indeed make people ill were being renewed to "eliminate the racially inferior parts of the city in order to 'reconquer' the city as a 'German city and native homeland' for the national community" (Münk, 1993: 293). Third, Nazis diagrammed the post-race city.

The post-race city is an imaginary of a new city where the "Jewish Question" has been solved once and for all because it is a city without Jews and other non-Aryan races; a city post-race as well as a city of racelessness but not because racialisation has ended but because it has been displaced or, better, it has been exteriorised, moved outside the city's borders. In 1939, Gottfried Feder sketched the post-race city in *Die Neue Stadt* (The new city). This diagram is not a "utopian fantasy of the future" (Feder, 1939: 476) to him but the result of "strictly scientific and practical treatises" (Feder, 1939: 479) on how to reorganise the German *Lebensraum* with new cities of 20,000 inhabitants. This is a technocratically determined city that grows "organically out of the social structure of the population" (Feder, 1939: 1). It has a concentric structure with a radial system of roads and two dominant axes. Centred and hierarchised: "a number of smallest communities are grouped together in one group and this group is reunited into groups of a higher order" (Feder, 1939: 467). Street communities are grouped into a subnucleus, subnuclei are grouped into a nucleus, and nuclei are grouped around the centre. In so doing, the "life of each district is clearly oriented towards its centre" (Feder, 1939: 19).

The National Socialist nature of the diagram reveals itself due to its "striking resemblance to the structure of the party" (Feder, 1939: 468). The post-race city is organised like the NSADP (National Socialist German Workers' Party). Both have a structure that is "quite organic" (Feder, 1939: 468). Indeed, "every city is an organism" (Feder, 1939: 18) and organic cohesion can only mean that certain "objects" and certain "animals"—that is racialised groups that are thingified and animalised—are either destroyed and killed or are firmly kept outside the city. The post-race city is a body: a body with organs where some bodies with no organs no longer cause a problem. It is the seeming opposite of a body under the conditions of viscerality. It is a *healthy* body, that is, it is a German body. The harmony of the new city equals the "the harmony and inner order of a well-grown being or other well-formed living being" (Feder, 1939: 2).

If, based on current post-racial claims, racism is only experienced in seemingly disconnected racial outbreaks avoiding structural differentiation, then in the post-race city racial outbreaks are not to be experienced at all. Neoliberal post-raciality amounts to "racisms without racism [as the] enigmatic condition of the circulation of racisms [...] without the terms to name it as such, to identify, comprehend, or condemn" (Goldberg, 2015: 81f). The anti-liberal post-race city amounts to *racism without racisms*: a city that is diagrammed as an everyday reminder, a continuous celebration of racism with all the Aryanised terms to name it as such, to identify, comprehend, and condemn the enemy. Jews are not an internal threat anymore. Yet, the organismic city that is centralised, hierarchised, and bounded requires the naming of an enemy for its self-definition and to prove its ongoing health. One potential threat of infection of the bounded post-race city, then, is the result of an "[u]ncontrolled immigration from the countryside", which, for Feder (1939) is one of the main reasons for the "decay of urban design and planning" (9). Another threat is due to "shameless land speculation and greed for profit" (Feder, 1939: 9).

While, in the early days of the NSDAP, Feder promoted economic theses of a rigid anti-capitalism, the diagram of the fascist city is, by and large, the diagram of the capitalist city, even if he makes an effort to eliminate excesses of capitalist urban planning (Petsch, 1976: 187). For example, in the post-race city, housing construction is separated from the capitalist order and interest-free loans for house construction are provided (Petsch, 1976: 191). To be sure, this is not a serious attempt to intervene in the capitalist order. Petsch (1976) alleges that "urban planning reveals the rational core of the irrational National Socialist conceptual apparatus: the urban structure and overall design consciously served to maintain existing economic and social structures" (190). But, typical of Nazis who speak against the "monstrosity of capitalism", the economic functions of the bourgeois in the capitalist production process are never attacked: the "bourgeois capitalist remains as the subject of the capitalist economy" (Marcuse, 2009: 7).

The post-race city is anti-bourgeois. It is also anti-liberal, because in the liberal city there is a "lack of authoritarian supervision of urban development and the construction industry" (Feder, 1939: 9). In the liberal city, "organic ties and rules for human leadership" of the German Middle Ages have been lost as the French Revolution "opened the door to the lack of barriers and so-called freedom" (Feder, 1939: 10). For Feder, liberalism and decay are almost synonymous. With regard to urban planning or, rather, the lack of it, building liberalism, that is the "freedom" to build whatever one wants wherever one wants, is a life-threatening infection. The anti-liberalist concept of freedom is a "new community idea" where the population does not form "'subjects' or masses of voters" but, instead, forms a "closed organism" (Feder, 1939: 18). The post-race city is a community, where "everyone lives through and for the other" and where the "freedom of the individual" has been replaced by the "sense of community" (Feder, 1939: 28):

> what time should be more favourable than the turning point brought about by National Socialism, and what field be more fertile and promising than the sacred soil of the Third Reich of all Germans – *Großdeutschlands* – which our Leader Adolf Hitler has awakened from millennia-old dreams to a power-conscious proud reality?
>
> (Feder, 1939: 479)

Liberal Urban Ecology (Berlin)

The post-race city is based on an anti-liberalist concept of freedom. And yet, it is in the liberal account of the Chicago School of Urban Ecology that we find the "classical" concentric urban diagram. Through organic and biological metaphors, an organisation that is "beyond the design of social actors" is imposed on the city and questions of social and spatial differences become "naturalised, i.e. depoliticised and socially reproduced" (Gassner, 2022: 792). In such a liberal concentric zone diagram, that is positivist and moralist, segregation is seen as a virtue. It is primarily through crossing borders, moving from one different neighbourhood with distinct habits, norms, and standards to another—crossing homogeneous "moral regions" (Park, 1969a: 128)—that the "freedom of cities" (Park, 1969b: 139) is experienced. "In such an account, forced migration and anti-Semitism are rendered invisible. [...] Here, freedom requires segregation" (Gassner, 2022: 793).

The Racist Garden City (Bristol)

In both the concentric city model of the Chicago School that embraces economic liberalism (c.f. Smith, 1988) and in the post-race city, freedom results from borders. Ebenezer Howard is commonly regarded as the pioneer of the garden city model. And yet, there are uncanny visual similarities between Howard's concentric diagram of the city and a version of the post-race city, which was developed as early as in the late 1890s by the anti-Semite Georg Fritsch. Fritsch believed in the absolute superiority of the Aryan race and became famous with *The Handbook of the Jewish Question* and the founding of the anti-Semitic journal *Der Hammer*. He was also an opponent of Einstein's relativity theory, was massively upset by the changes brought by industrialisation, and diagrammed his version for the post-race city—*The City of the Future* (Die Stadt der Zukunft)—in 1896, two years before Howard published *Garden Cities of Tomorrow*. At one point, he accused Howard of plagiarism. Is the "true" father of the garden city an anti-Semite?

Fritsch's post-race city shows visual similarities to Howard's garden city but the two operate fundamentally differently. While Howard diagrams an alternative to industrial capitalism, Fritsch's emphasis is on social inequality and authoritarian government. Howard's socio-political and reformist aspirations are anti-urban resentments in Fritsch's work. Still, while for Howard, land in the garden city has to be in common ownership in order to replace the capitalist competitive society with a society with a cooperative basis (Petsch, 1976: 188), at one point Fritsch (1896) also claims that land "must be and remain the property of the municipality; it is only to be leased to citizens for longer periods of development" (15).

Visual similarities remain in their imaginaries of cities in the future. Arguably, there are important lessons to learn for "the democratic character of all today's cosmopolitan imaginings" (Gilroy, 2000: 2). It was in his early years in Vienna that Hitler "had ceased to be a weak-kneed cosmopolitan and [became] an anti-Semite" (Hitler, 1969: 59). Fritsch (1896) voices his anti-urban resentments in a clear enough way: "Cities especially large cities are now regarded as unhealthy excesses of civilization, as 'water heads' and 'plague bumps of culture'. Kingsley even called them 'pig stables of culture' and unfortunately with a certain right" (Fritsch, 1896: 5). Cities are unhealthy because "what is missing above all is the 'inner order'" (Fritsch, 1896: 4). And it is contended that an inner order can only be achieved when the Jewish problem has been solved once and for all.

Fritsch's organisation of the city body is the result of strict zoning. His diagram is a concentric one with zone 1 (centre) as the location for monumental public buildings; zone 2 for monumental villas; zone 3 for better residential buildings; zone 4 for residential and commercial buildings; zone 5 for working-class housing and small workshops; zone 6 for factories, construction yards, and storage areas; and finally, zone 7 for nurseries and rented gardens. Shops, schools, post offices, theatres, and other amenities are located in all different zones, making sure that the exchange between different social classes is minimised as much as possible. Not unlike in Park's liberal city, order means that "the same follows the same, the relative mates with the relative" (Fritsch, 1896: 8). Harmoniously grouping "individual parts according to their internal relationships" (Fritsch, 1896: 5) is based on the assumption that difference, different social groups (race is not an issue anymore) means the disintegration of communities: a particular but not uncommon understanding of communities (and social cohesion) as resulting from grouping sameness, that is separating difference.

The organic structure of Fritsch's city results from the "division into socio-economic classes and the ranking of different strata of society" (Petsch, 1976: 190). The organicism of this post-race city requires the visibility of difference, but not (unnecessary) contact between difference. This is the diagram of a city for "people with 'strong physical and moral health'" (Münk, 1993: 173). Here, strict zoning is the spatiality of class and social status. Freedom is to be found within one's zone, knowing one's place in the city, reinforced by the strict spatial organisation that is a constant reminder of one's position, that is one's location vis-à-vis other locations and therefore vis-à-vis other socio-economic classes. So, spatial borders need to be transgressed not necessarily physically but in one's imagination to cement one's position within the city as an organism. What keeps the urban diagram in shape, what fixes people's positions within this post-race city, is not only the imagination of another zone but also the recollection and imagination of racist events as potential attacks on the organism's health; and all that in the absence of experiencing racisms.

Aesthetics after Race (Bristol)

The 1930s post-race city is an imaginary of a city after experiential racisms. Post-raciality nowadays is "the illusion that the dream of the nonracial has already been realized" (Goldberg, 2015: 180). Both are characterised by a disconnection between structural and institutional racism on the one hand and the experience of racisms on an everyday basis on the other. While post-raciality means racisms without racism, the post-race city means racism without racisms. The post-race city is based on the concept of race. Yet, due to eliminationist over exploitative racism within its bounds, racisms (again, within bounds) have been made to stop. In this chapter, I also suggest that the organismic city, which the post-race city is one particular imaginary manifestation of, can be fascist or liberal. The organismic city has a "unity of composition" (Smith, 2017). As a composition, it is centralised, hierarchised, and self-directed. It is bounded. In the fascist anti-liberalist city, freedom is within borders (organic social cohesion inwardly directed: the folk) and as a result of borders (organic social cohesion outwardly directed: against the enemy). In the liberal city, freedom is a result of crossing borders. In both the fascist and liberal cities, freedom is linked to, and relies on, borders.

An urban aesthetic after race has, for sure, to find ways of engaging with as well as against an aesthetic theory that was developed in the eighteenth-century European Enlightenment as a project that explored conditions of possibility of universal human subjecthood and, in so doing, created the civil, liberal, and anti-revolutionary subject (Lloyd, 2019; Jackson, 2016; Roelofs, 2015; Brigstocke and Gassner, 2021). In this chapter, I position aesthetics as a relationship between experiences and imaginaries. I make a case against an *unbounded aesthetic*, that is simply the imagination of a world without borders. Such an aesthetic renders histories and processes of borderisation invisible. Instead, an urban aesthetic after race involves a continuous engagement with and an un-thinking and un-doing of borders: an *aesthetic of un-bordering*. The difference between an unbounded aesthetic and an aesthetic of un-bordering relates to Goldberg's (2015) distinction between racial non-racialism, which "seeks to sidestep historical legacies of racial arrangement and injustice" and anti-racism, which "seeks to critically address and redress the impacts of those legacies" (20). An aesthetic of un-bordering works with and against processes of borderisation; it works with and against the concept of race: it is one that is genealogical in order to imagine alternative futures. It is for good reason, then, that the banner I read in Berlin stated: BORDERS KILL: SOLIDARITY EVERYWHERE. It did not read: NO BORDERS: SOLIDARITY EVERYWHERE, precisely to bind imaginaries to racialised lifeworlds.

Note

1 https://www.facebook.com/kfetisch/.

References

Arendt, H. (1978). *The Life of the Mind: One/thinking*. San Diego: Harcourt.
Arendt, H. (1992). *Lectures on Kant's Political Philosophy*. Chicago: The University of Chicago Press.
Benjamin, W. (2006a). On the Concept of History. In *Selected Writings. Volume 4, 1938–1940*, edited by H. Eiland and M. W. Jennings. Cambridge, MA: The Belknap Press of Harvard University Press, pp. 389–400.
Benjamin, W. (2006b). The Work of Art in the Age of its Technical Reproducibility (Third edition). In *Selected Writings. Volume 4, 1938–1940*, edited by H. Eiland and M. W. Jennings. Cambridge, MA: The Belknap Press of Harvard University Press, pp. 251–283.
Bonta, M. and Protevi, J. (2006). *Deleuze and Geophilosophy: A Guide and Glossary*. Edinburgh: Edinburgh University Press.
Brigstocke, J. and Gassner, G. (2021). Materiality, race, and speculative aesthetics, *GeoHumanities*, 7(2), pp. 359–369. DOI: 10.1080/2373566X.2021.1977163.
Burleigh, M. (2000). *The Third Reich: A New History*. London: Pan Macmillan.
Deleuze, G. (2015). *Foucault*. London: Bloomsbury.
Deleuze, G. and Guattari, F. (2016). *A Thousand Plateaus*. London: Bloomsbury.
Feder, G. (1939). *Die neue Stadt: Versuch der Begründung einer neuen Stadtplanungskunst aus der sozialen Struktur der Bevölkerung*. Berlin: Springer Verlag.
Foucault, M. (2009). Preface. In *Anti-Oedipus*, edited by G. Deleuze and F. Guattari. London: Bloomsbury, pp. xiii–xvi.

Fritsch, T. (1896). *Die Stadt der Zukunft*. Leipzig: Verlag von Theodor Fritsch.
Gassner, G. (2020). *Ruined Skylines: Aesthetics, Politics and London's Towering Cityscape*. London: Routledge.
Gassner, G. (2021). Drawing as an ethico-political practice. *GeoHumanities*, 7(2), pp. 441–454.
Gassner, G. (2022). Spiral movement: Writing with fascism and urban violence. *The Sociological Review*, 70(4), pp. 786–809. DOI: 10.1177/00380261221106526.
Gassner, G. (2023). Revolutionärer Antifaschismus als Stadtplan: Gewalträume, Freiräume und Traumräume. Sub\urban, 11(3/4), pp. 199–231. DOI: 10.36900/suburban.v11i3/4.862.
Gilroy, P. (2000). *Against Race: Imagining Political Culture Beyond the Color Line*. Cambridge, MA: The Belknap Press of Harvard University Press.
Goldberg, D. T. (2015). *Are We All Postracial Yet?* Cambridge: Polity Press.
Hitler, A. (1969). *Mein Kampf*. London: Hutchinson.
Jackson, M. (2016). Aesthetics, politics, and attunement: on some questions brought by alterity and ontology. *GeoHumanities*, 2(1), pp. 8–23. DOI: 10.1080/2373566X.2016.1165076.
Lennard, N. (2019). *Being Numerous: Essays on Non-Fascist Life*. London: Verso.
Levinas, E. (1990). Reflections on the philosophy of Hitlerism. *Critical Inquiry*, 17(Autumn 1990), pp. 62–71.
Lloyd, D. (2019). *Under Representation: The Racial Regime of Aesthetics*. New York: Fordham University Press.
Marcuse, H. (2009). *Negations: Essays in Critical Theory*. Minneapolis: Mayfly Books.
Mbembe, A. (2019). *Necropolitics*. Durham: Duke University Press.
Münk, D. (1993). *Die Organisation des Raumes im Nationalsozialismus: Eine soziologische Untersuchung ideologisch fundierter Leitbilder in Architektur, Städtebau und Raumplanung des Dritten Reiches*. Bonn: Pahl-Rugenstein Verlag.
Neocleous, M. (1997). *Fascism*. Maidenhead: Open University Press.
Park, R. (1969a). The City. In: *Classical Essays on the Culture of Cities*, edited by R. Sennett. Des Moines, IA: Meredith Corporation, pp. 91–130.
Park, R. (1969b). Human migration and the marginal man. In *Classical Essays on the Culture of Cities*, edited by R. Sennett. Des Moines, IA: Meredith Corporation, pp. 131–142.
Petsch, J. (1976). *Baukunst und Stadtplanung im Dritten Reich: Herleitung/Bestandsaufnahme/Entwicklung/Nachfolge*. München: Carl Hanser Verlag.
Roelofs, M. (2015). *The Cultural Promise of the Aesthetic*. London: Bloomsbury.
Sennett, R. (2011). Boundaries and borders. In *Living in the Endless City*, edited by R. Burdett and D. Sudjic. Phaidon, pp. 324–331.
Smith, D. (1988). *The Chicago School: A Liberal Critique of Capitalism*. Basingstoke: Macmillan.
Smith, D. (2017). What is a body without organs? Machine and organism in Deleuze and Guattari. *Continental Philosophy Review*, 51, pp. 95–110. https://doi.org/10.1007/s11007-016-9406-0
Tyner, A. H. (2017). Action, judgment and imagination in Hannah Arendt's thought. *Political Research Quarterly*, 70(3), pp. 523–534.
Valluvan, S. (2016). What is 'post-race' and what does it reveal about contemporary racisms? *Ethnic and Racial Studies*, 39(13), pp. 2241–2251.

7 The Walls of Berlin and Cairo

A Spatial Mnemonic Model

Taher Abdel-Ghani, Yara Mohamed, and Amr Ibrahim

Introduction

In one scene in Wim Wenders's 1987 German film *Wings of Desire*, the angel Damiel is persuaded to experience life for the first time. He bleeds, sees colour around him, tastes food, and drinks coffee (Wenders, 1987). In doing so, he creates unique mental images of the new physical environment he has immersed himself in, a response to "affective intensities of life in shared public spaces" (Latham and McCormack, 2009; McCormack, 2013; Pink, 2008 as cited in Lobo, 2019: 1). Through the action of seeing and touching, Damiel's body became part of the city and gradually recognises the city as being a part of him, establishing a "holistic conception and representation of the city and its spaces" (Tiwari, 2010: 137).

Gilles Deleuze addresses this complex dimension of spatial connectivity as "the haptic", which relates to touch and bodily experience, signifying a deep, embodied engagement that is of a sensory nature, rather than focusing on more cognitive or representational aspects (Deleuze, 1981; Yoshitaka, 2013). Deleuze's theory of the haptic posits that an image or object fuses colour, odour, sound, and taste, capturing the emotions and experiences of users. The chapter argues that such an approach enables a consideration for the emotional and visceral responses to certain images—such as graffiti—raising questions of memory and how it affects, and is affected by, materiality (Totaro, 2002; Deleuze, 1985). Hence, memory becomes an integral consideration in thinking about emotions, affects, and material factors over space.

As Damiel comes to realise his own existence in the real world, he walks by the Berlin Wall, decorated with French artist Thierry Noir's distinctive head paintings. Noir himself plays the man on the ladder painting the wall and waving to Damiel. Having worked on three miles of the wall, Noir, who was Berlin Wall's first graffiti artist, emphasised the historical significance of the wall as a reminder to the upcoming generations of the casualties caused by the wall (Burke, 2009). In an interview with Street Art London (2013), Noir considered his act of painting the wall as both a revolutionary act against its oppressiveness, and as a mutation of nature and culture. Noir's unique painting style of human heads resulted from his struggle to answer questions targeted to him

DOI: 10.4324/9781003335993-7

whilst he was painting; hence, he developed a style of painting where it was possible to talk and paint at the same time. Eventually, his paintings would go on to become a symbol of the new freedom in Europe.

With the works of other revolutionary artists, such as Brigit Kinder and Keith Haring, to which we later turn, the art on the Berlin Wall became an iconic symbol of the Cold War division. By 1989, the wall illustrated the sentiments of Berlin's population and has since redefined urban aesthetics through its embedded questions of memory and materiality, which themselves have become woven into intellectual writings about spatial and temporal aspects of the built environment. The Berlin Wall is now perceived as more than a mere object standing at a specific site; given the remnants and evocations of World War II it embodies the paradox of being a static structure deeply entangled in ongoing, dynamic social and historical processes, functioning as a generative but divisive force of unresolved tensions, a "creative aggressive agent of unfinished business" (Bach, 2016: 59). Jones (2014: np) describes it aptly: "The Wall froze time. It prevented Berlin from becoming a neat post-war city, kept it semi-ruined and shadowed by its past as well as present".

Fast-forward to 2011 in Cairo, where similar pseudo-artistic, pseudo-political events occurred. The Mohamed Mahmoud Wall, adjacent to Tahrir Square, witnessed a revolutionary eruption of graffiti artists voicing against security forces. The wall has since become the subject of several scholarly debates on the power of space and the space of power (Bayat, 2010; Mitchell, 2003; Harvey, 2012; Rabat, 2011). On the ground, it served as a centre of attention for anyone interested in learning about socio-political struggle. The Mohamed Mahmoud Wall reintroduced a cultural phenomenon that was once hindered under the Mubarak regime. Similar to the Berlin Wall, the graffiti work on the Mohamed Mahmoud Wall has become an integral part of the formation of public space imagery, constituting an important section of urban cultural memory. At the time, it was a first-of-a-kind experience for the Egyptian masses to witness a unique dialogue between public art and the space it inhabited (García-Doménech, 2015). Hence, the artwork embodied a tangible collective consciousness that is not static in nature but serves as a "fluid reconstruction that feeds into current needs as well as future imagination" (Awad, 2017: 236).

Revisiting the historical significance of each wall, the chapter highlights how graffiti captures and contributes to the city's aesthetic character while serving as a catalyst for civic movements. It adopts Tursić's approach (2019), considering space and time as the basic pillars for supporting knowledge regarding urban aesthetics. This chapter focuses on the mnemonic spatial structure through which aesthetic experience occurs. In the case of Berlin, the Wall symbolised world division and the oppression of individuals by the state (Ebeling, 2019). Conversely, the Mohamed Mahmoud Wall in Cairo underwent a vivid transformation from being a mere concrete wall surrounding the American University in Cairo to an iconic mural canvas for the public against social, political, and economic inequality. The significance of this wall was further exemplified in the months following the revolution, during the clashes between protestors and security forces from 19th until 24th November 2011. The wall provided opportunities for artistic, civil rights, and youth coalition movements that were continuously obstructed by authorities.

The common interest in both cases is the co-evolution of the walls with the people interacting with them and the political contexts such interactions arise from. There is an emphasis on the notion of time as relating to the "virtual" dimension accompanying the walls (Latham and McCormack, 2004; (Massumi 2022). By virtual, here we evoke Deleuze's (Deleuze, 1981, Deleuze, 1985) understanding, where in contrast to the actual as concrete and realised entities, the virtual represents a kind of potentiality that is not yet actual but is nonetheless present in the sense that it exists as a potential. In this sense, the walls project collective memory that sow seeds for the creation process of urban spaces and city cultures. They can be considered agents of historical moments and representations of a once-emerged socio-spatial identity that shapes the image of the city (Loughran et al., 2015). As embodiments of memory, the walls of Berlin and Cairo are what we call "mnemonic spatial models" that (might) work to establish a social and cultural "truth" (Tursić, 2019).

In its basic technical definition, "mnemonic" refers to an aiding technique useful for retrieving information from human memory. A common example is children's learning of mathematical multiplications through a certain musical rhythm. Studies on contemporary urbanism have paralleled cities' mnemonic spatial behaviour to that of the human mind, in which urban objects contribute to the manifestation of memory. This chapter explores the walls of Berlin and Cairo as spatial mnemonic objects that encapsulate characteristics and images of both cities and serve to provoke collective and individual memories. It examines how their transformational aesthetic qualities are the result of their respective political circumstances, the temporal evolution of urban spaces built upon social upheaval, and the visualisation of memory and value within the fabric of the city through symbols and objects (Popczyk, 2015).

Literature Review: The Aesthetics of Spatial Memory

This section looks at previous writings on memory and aesthetics. It outlines the approach to studying aesthetics adopted in this chapter and its incorporation into thinking around time and cultural space. Additionally, the literature review explores the theory of "the haptic" and considers memory and aesthetics as related ideas. To reinforce this, the second part of the literature review critically considers examples of graffiti from different countries, such as Greece and Russia and highlights mnemonic memory as an often visual, augmentative language that reflects the social and ecological systems that comprise urban spaces.

In 2006, fieldwork conducted between the universities of Seville and Southampton explored the geological and physical properties of natural substances from the Neolithic and Copper Ages (Sanjuán and Wheatley, 2010). Their findings revealed iconic monuments that "established highly patterned relationships between places within the landscape, conveying notions of time and cultural memory" (Sanjuán and Wheatley, 2010: 22). This research, which examined the role of natural substances, landscape forms, and funerary monuments, emphasised how objects can articulate a collective sense of memory and place. Such a theme was previously highlighted in Gaston Bachelard's *The Poetics of Space* (Bachelard, 1964), for whom space is seen as a repository for individuals' memories and experiences which become accumulated over time (Lillios and Tsamis, 2010; Bachelard, 1964). Within Bachelard's framework emerged a continuous loop between narratives of the past and the events of the present, in the sense of appreciating the role of history in shaping and enabling the present to come into being.

The idea of memory and its imprint on socio-spatial aesthetics has been addressed within Western thought. Urban memory, as some scholars call it (c.f. Wang et al., 2016; Crinson, 2005; Othman et al., 2013; Postalcy et al., 2006), is understood as being created out of individuals' imaginaries within a place and its cultural and historical background. In this sense, memory, reinforced with spatial and temporal anchors, is akin to the concept of "*Rikyū* Grey", also known as the "philosophy of grey" (Yang, 2019). It was first introduced in a 1977 special issue of *The Japan Architect* referring to the ambivalence of Japan's open spaces. Whereas in the West grey is a mix of black and white, in Japan, Rikyu grey is a blend of red, blue, yellow, and white, with the exact formulation affecting the shade of grey produced. It can nonetheless be seen as colourless, where the vibrant red, blue, yellow, and whites are in effect cancelled out through their blending, thus achieving continuity and coexistence. Just as Rikyū Grey captures multiplicity, ambiguity, and the coexistence of meanings, spaces of urban memory embody the layered interplay of past, present, and future. These spaces oscillate between the "imagination of the past and the foresight for future events" (Boyer and Wertsch, 2009: 18), much like the subtle tones of Rikyū Grey invite an appreciation of complexity, transience, and coexistence. Thus, Rikyū Grey serves as a conceptual metaphor for urban memory.

This multiplicity and ambiguity of meaning became a subject of concern regarding the study of memory. This is because memory, in all its multiplicities, lies at the core of rapid social transformation and people's search for a past within the rubble of current social, political, and economic forces (Hoelscher and Alderman, 2004; Harvey, 1989). One therefore requires a multi-layered interpretation of space and the embeddedness and evocation of memory within it (Zimmerman, 2008; Koetter, 1980). Authors understanding space in this way have highlighted how the integration of art into urban spaces—such as public or street art—creates a network of relationships involving space, time, motion, surfaces, and users. Doing so redefines social experiences and reshapes the cultural dynamics of these spaces. Abdel-Ghani (2020), for instance, references Karl Kropf's theory of phylogenetic change—a concept from biology describing the evolutionary history of a group of organisms—to analyse the impact of cinematic interventions in Tahrir Square during the January 2011 protests in Cairo. Abdel-Ghani (2020) highlights how the collective mobilisation of different elements of the city (film, art, and motion) contributed to redefining the built environment.

This chapter explores how memory relates to space and time, emphasising the differences between Western and Eastern paradigms of aesthetics. In doing so, it implicitly moves between two approaches to studying spatial aesthetics, as outlined by Alle (2012): *single space and complex space*. By single space, Alle refers to the cognitive perception of space as a vertical process, focusing on physical elements—such as lakes or churches—that transform a space into a meaningful place. Complex space, on the other hand, sees space as a horizontal process, emphasising the dynamic social and spatial networks that create transitory spaces. Movement, performance, and human activity play a central role in shaping these spaces.

To the end of understanding movement, we might look to the work of McCormack (2008) who considers the movement of bodies in a twofold manner: first, the physical movement (e.g. walking, running, and falling), and second, the affective, imaginative, kinaesthetic, social, cultural, and political dimensions of a body's movement. In this sense, McCormack highlights how movement and performance within spaces can generate places. This distinction between space and place is a broader theme in human geography and cultural theory (Massey, 2005). McCormack's (2008) approach, drawing on non-representational theory, emphasises how meaning—and indeed place—continuously emerges through the dynamic movements of both human and non-human bodies, rather than relying on reductive or static representations (Harrison, 2000; Thrift, 2008).

This consideration for movement is also present in works on art. Biggs (2009), for instance, contextualises contemporary art in public space by examining how movement through these spaces can create a sense of place. This dynamic, organic relationship between artists, materials, and space is what some have argued actually generates the artwork, which Bolt (2010) refers to as *methexis*. *Methexis*, in this context, refers to a participatory process where audiences actively engage in the creation of the artwork or performance, rather than being passive observers. This participatory approach reflects a networked relationship between humans, materials, and space. This approach also challenges traditional classifications of aesthetics, such as those proposed by Bobaroğlu (2005), who contrasts Eastern and Western views. According to Bobaroğlu (2005), Western aesthetics prioritise form, while Eastern aesthetics emphasise content and essence.[1] Instead, foregrounding a non-representational approach to aesthetics that is attuned to movement and methexis prioritises uncertainty and the constant flux of cultural identity and history.

In this regard, the authors of this chapter second Latham and McCormack's point that if we are to speak of urban experience—as encompassing the "relations between human and nonhuman forms of life"—then we ought not to think about this as dictated by some "central controlling or decision-making unit" (Latham and McCormack, 2004: 707). Instead, making sense of urban experience requires questioning immaterial entities (e.g. affects, events, or ideas) and their relation to material qualities (e.g. objects or bodies), which are interwoven in assemblages (Guattari, 1995). Importantly, these relations are always changing—they are moving—and so must be understood as emergent.

Introducing the idea of memory as aesthetics builds upon this idea by highlighting how both material and immaterial elements coalesce to shape urban life. The focus on both material and immaterial factors is owing to how memory can entwine with physical forms such as cathedrals and resting places, but also with the immaterial, such as celebrations and rituals (Said, 2000; Hoelscher and Alderman, 2004). Memory, embedded within social contexts, generates performative urban landscapes, in that cultural practices, identities, and histories are continuously enacted and reinterpreted, shaping the spaces in which they occur.

In thinking about artworks and memories, it might be argued that this chapter considers representations of the past. However, according to Thrift (2008), representation is confined to a world that already exists—one in which, as Foucault argues, "order, truth, and the subject are paramount" (Foucault, 1971; Boyd, 2017: 28). Given the aforementioned focus on movement, we do not seek to render static graffiti or memory in this regard. The authors of this chapter instead theorise memory as part of a wider system that involves human performance, that is movement and touch, which non-representational theory is well attuned to in its focus on how "life 'takes place' through movement, intensities, and encounters" (Boyd, 2017: 29; Lorimer, 2005). Memory, then, becomes associated with performance, and both redefine the aesthetics of the city. Social memory becomes mnemonic, in that it functions as a collective, shared system for storing and recalling memories, not just through verbal recounting, but through various material and immaterial forms of cultural expression (e.g. rituals, monuments, art, public spaces, and even graffiti). Memory is therefore intricately folded into what Thrift (2008: 2) refers to as "the geography of what happens". This chapter explores street graffiti as part of urban aesthetics and in relation to memory and the multiplicity of relations existing within urban space. Similar to Damiel's haptic experience with reality, it argues that people's participation and interaction with street graffiti evoke their sense of culture, identity, and socio-spatial status.

Evoking Mnemonic Cultural Memory: Examining Graffiti

Graffiti is temporary, and city authorities, especially totalitarian governments, have often perceived it as an act of vandalism leading to its removal. Graffiti is seldom intended to be permanent and is often painted over within a few days or hours. However, grassroots artistic methods have attempted to preserve graffiti virtually. In 2021, the Russian artist duo Yav expressed their opposition to the Russian government and the laws restricting freedom of speech through a street art painting with the word "*Терпение*" (Patience) (Brosius, 2021). Although the street art was physically displayed for only nine hours, it remains accessible in the same location through an augmented reality (AR) mobile app. This represents a process of "un-walling the wall [that] becomes the undoing of law" (Weizman, 2007: 210). The wall itself, therefore, takes on a "palindromic linguistic structure", illustrating the deep interconnection between the built environment and legal frameworks (Weizman, 2007: 209; Davies, 2017).

Graffiti provides alternative perspectives on subcultural movements, reflecting the self-expressive efforts of artists who adopt alter egos and their connection to cities. Through the spatial narratives and actions of the artists, graffiti shapes how people perceive history, economy, and place. This process emerges from the intangible dialogue between the artwork, artist, and viewer (Hansen and Flynn, 2015; Fransberg et al., 2021). Graffiti's integration into urban spaces has significantly influenced how cities are visually represented and experienced, highlighting the impact of its coexistence on urban imagery.

Taking Greece as an example, Johannis (2011) examined the grassroots, social and political roles of graffiti in revolutionising the masses against the monotonous architectural landscape of Greece that resulted from the intervention of multinational companies within the urban fabric. For Johannis (2011: 21), graffiti often symbolised sovereignty of the urban environment and served as a powerful visual expressive language in its rejection of (i) global consumerism constituted in the form of "shady propaganda which advocated the violent distortion of public spaces" in favour of commercial interests; (ii) the new social and political imagery of Greece of the 1990s and early 2000s which witnessed the deterioration of cultural and educational programs; and (iii) the emergence of a monotonous and lifeless urban landscape with few green areas and limited sites of historical and cultural significance.

Another example is put forward by Petrov et al. (2019) with the Russian city Almetyevsk, a small industrial city that overcame its post-Soviet identity crisis through the public art project "Tales of the Golden Apples". The five-year project tackled the historical and cultural codes of Almetyevsk, created public art that contributed to the transformation of the built environment, and established educational programs for locals and newcomers. Take, for instance, "The Myth of the Golden Bird" painted by the Greek artist Fikos which depicts a local traditional narrative as an attempt to revive the city's folkloric identity. Using the visual language of graffiti, text, and folkloric imagery, the project as a whole successfully bridged the relationship between artists and architects, creating a shared territorial space for local inhabitants. This collaboration contributed to the development of a distinct local cultural identity (Petrov et al., 2019).

Graffiti, then, weaves together the spatial and temporal, telling the story of a city and its society. In many cases, graffiti brings together and narrates social and political events, civic voices, and feelings of collectivism and micro-resistance. The actual graffiti is not confined to the visual or representational but is instead a haptic endeavour. It is haptic because the act of moving past or around the artworks allows people to engage with the art in a more embodied way—going beyond the purely visual or intellectual to relate to a more bodily and sensory experience of interacting with the world. The proximity to the art and the ability to view it from different angles whilst journeying through urban environments creates a haptic relationship between the viewer and the work. As Bowen (2013) puts it, such artworks are "a spatializing practice that claims spaces, makes places, and illuminates margins and borderlands". The chapter will now explore such a phenomenon in two cities, Cairo and Berlin, where each has gone through a time of political unrest and where art has played a significant role in representing the social response.

The Walls of Cairo and Berlin

During the protests of January 2011 in Cairo, specifically in Tahrir Square which is the central iconic space for civic resistance.[2] Seven interviews were conducted with Egyptian graffiti artists by either scholars or local newspapers about the significance of their works' intervention in the street. Among them were Keizer, Alaa Awad, El-Teneen, Ammar Abo-Bakr, and the most prominent artist, Ganzeer, who drew similarities between the graffiti resistance process in Cairo to the one in Berlin in 1989 before the fall of the wall. Ganzeer co-organised the *Martyr Mural Project* in honour of a 16-year-old killed by security forces.

Such similarity captivated the attention of the authors, questioning the common ground that can exist between the two cases of Berlin and Cairo. In both, public paintings had a significant aesthetic impact, transforming the streets while simultaneously revealing the broader socio-political contexts conveyed by the artists. Additionally, these illustrations were also ones of resistance that took aim at the nation-state and its operations.

In Berlin, the image of the modern city in the 1950s and 1960s that adopted American post-war housing policies was replaced with paintings of dissent and anti-governmental efforts, while in Cairo, the walls of public institutions were transformed into dynamic canvases objecting to the control of the regime over public spaces. In both cases, all traces of civic resistance, whether visuals or objects, were later replaced with central, national visions to ensure the state's superiority. On the one hand, Berlin's governmental plans were to selectively restore some of the artworks—removing "unwanted" graffiti that had emerged in the intervening years—remunerating artists for their recreation as part of a gentrification process (which we shortly detail), while on the other, Cairo's authorities have replaced the martyr's memorial with an obelisk that many consider to be an off-cultural object (Burke, 2009). Though short-lived, from 2011 to 2013, Cairo's graffiti left a significant imprint on society's interpretation of the city.

Studies have considered the interaction between memory and the walls of Berlin and Cairo, with graffiti as the focal point, however, attention is yet to turn to exploring graffiti's interaction with those that consume it, focusing on bodily and sensory engagements with public space (Lobo, 2019). Both the Berlin Wall and the Mohamed Mahmoud Wall are often associated with traumatic pasts. Graffiti mobilised the preservation of memory to cope with problematic and conflicting pasts, as well as to prevent similar events from occurring in the future (Pinto, 2017).

Berlin's Quest for Revival

Berlin's 1989 "Fall of the Wall" was not just a signification of the end of a horrific era, but also the beginning of a new quest for European revival, which witnessed the fall of communism in Central and Eastern Europe, an end to the Cold War era, and the German reunification the following year. Following the fall of the Wall, Bach (2016) stated that three key sites for commemoration emerged: the Berlin Wall Memorial, Checkpoint Charlie, and the most prominent site, the East Side Gallery. The visual memorials on the Wall had already commenced before 1989 on the Western side. Artists were already commemorating the dead and breaking the wall through either visual linguistics, as prominent graffiti artists Thierry Noir and Keith Haring daringly did, or through sound installations, such as American artist Ed Kienholz's work in West Berlin creating sound installations that explored the dark history of Nazi Germany (Mesch, 2008). Consider, for instance, German artist Birgit Kinder's mural "Trabant" painted at the East Side Gallery. It is a depiction of a Trabant car, the car driven by most Germans at the time, smashing through the wall. Between February and September 1990, 118 artists from 21 countries created around 106 artworks on the remnants of the Wall and the East Side Gallery (Caldas, 2015). By November 1991, a portion of the wall was designated a national monument, and five years later, Iranian artist Kani Alavi founded the East Side Gallery e.V. artists' initiative to oversee the preservation of their artworks (Prior, 2019).

It is noteworthy that several segments of the Wall are scattered in exhibitions around the world, and little has survived in its original hometown of Berlin. One major reason for this is the desire of some Berliners to erase all that is associated with the traumatic events related to the Wall, which the artwork decorating the Wall depicts. This argument was reinforced by Dr. Volker Hassemer, the Senator of West Berlin responsible for urban development before the fall of the Wall, who claimed that the country should recognise the existence of the remnants as witnesses of the past (Schmidt and Klausmeier, 2004).

The longing for the preservation of the collective memory, which today seems in many ways to have been accomplished, was at the time a challenging task. According to Bach (2016), Berlin struggled to overcome the division, and the state had no plans for a Wall memorial. Reform planning for the city involved erasing any traces of the wall to establish a new era of unification. Yet, with the ongoing property disputes at the time that delayed any new construction project, several segments of the wall were left abandoned. By 2009, following years of being exposed to vandalism, the original Berlin Wall artworks were subject to being repainted, which most of the artists refused. As Brigit Kinder puts it, "the original images were painted to capture the spirit of the times. If we don't preserve that, we'll soon have a Disneyland where everyone just paints what he [sic] likes" (Overdorf, 2016: np).

Artists were expressive about the political and cultural contexts of Berlin at the time. Institutions invited them to give graffiti a powerful voice of resistance to the "ridiculousness" of the Wall, which often clashed with civic authorities

(Pugh, 2015). Looking at some of the artworks and the themes embodied, one can notice messages of protest being communicated in several forms and colours. French artist Thierry Noir, considered to be the first artist to paint the Wall, narrates his experiences with painting the East Side Gallery (Jones, 2014). Inspired by Lou Reed's 1973 song *Berlin*, Noir's revolutionary act to paint the wall was a soft response to the psychological and physical violence that resulted from the division of the city. Among his early paintings on the Wall were dinosaurs as a representation of the mutation of nature, while he metaphorically referred to himself and the Wall as mutations of culture (Jones, 2014). Such mutation is portrayed in Keith Haring's interlocking chains of human figures representing the unity of the people against the idea of the wall. Haring's work was inspired by the hip-hop culture of 1970s New York, a deviant subcultural genre that Haring integrated within the Berlin Wall (Brooke, 2007).

Interpreting this countercultural post-modern visual movement that intervened in Berlin, Miller (2002) identifies two main themes of graffiti associated with Haring and Noir's work: (i) constructing an identity that opposes the state and consumer culture and (ii) establishing the idea of resistance through cultural production. The paintings on the Berlin Wall were also a countermovement against the prefabricated concrete structures that took over the planned fabric of Berlin at the time which was largely comprised of high-rise apartment complexes (Pugh, 2015). Architectural styles of socialist realist structures attempted to project a luxurious and aspirational image of life to the West, in stark contrast to the conditions faced by many citizens.

Cairo's Visual Conflict

Twenty-two years after the fall of the Berlin Wall, mass protests took place in different regions of Cairo. This unexpected turn of events—part of the so-called Arab Spring movement—encompassed demands for greater political freedoms and the resignation of President Hosni Mubarak, who had ruled Egypt for nearly 30 years. Tahrir Square, in central Cairo, was central to this, and witnessed struggles between civic voices and security bodies over the control of public space, to the point that the square became occupied with its own self-organising movement. The January 2011 revolution has been referred to by some as an urban revolution (Nagati and Stryker, 2013; Abaza, 2013; Awad, 2017). Unlike, for example, the French or the Russian revolutions which saw the overthrowing of imperial governments by the local parties, the Egyptian revolution's centre of attention was Tahrir Square as the main spatial protagonist that was under the spotlight for a relatively short span, from 2011 to 2013, compared to the Berlin Wall. The key events included the overthrowing of Mubarak in February 2011, the assignment of the Supreme Council of the Armed Forces for more than a year as a transitional stage, the Muslim Brotherhood's take-over of the presidency until June 2013, and eventually, their disposition by the masses and the army, all of which caused significant spatial transformations in the imagery of the city of Cairo and its people (Awad, 2017).

The most striking aspect of the revolution was its ability to foster a reimagined public culture, epitomised by Tahrir Square becoming a "city within a city". It served as a microcosm of the ideals and aspirations of the revolution, where people collectively redefined spaces for civic engagement, collaboration, and solidarity. Different civic functions—such as food vending, medical support, selling flags, and outdoor haircut salons—were divided into zones across the square. The public space was a stage for exhibiting and performing acts of civic resistance against Mubarak's dictatorship. Abaza (2013) highlights the war zones that have been created through the erection of barricades, barbed wires, and controlled areas of security forces to control the anger of the masses. Mohamed Mahmoud Street was the first street to witness such war zones, with the death of more than 40 protestors by the police.

In an act of artistic resistance, graffiti artists painted over the concrete blocks installed by the military to block protestors' movement, creating a striking depiction of the street that lay hidden behind them. Rendered in a photorealistic style, the artwork imagined the non-presence of the concrete blockade, symbolising the freedom that was obstructed. This street, also known as *Sharei' Uyuun Al-Hurriyyah* (The Street of the Eyes of Freedom), has become an iconic space for commemorating martyrs who witnessed some of the most violent moments in November and December 2011 under the rule of the Armed Supreme Council. The Mohamed Mahmoud Wall was a continuous, yet short-lived, conflict between the artists' creative voice and the military junta.

The aesthetic and political significance of the graffiti artwork of Mohamed Mahmoud Wall is associated with the variety of cultural and historical themes depicted through the paintings. Some of them are re-portrayals of the Ancient Egyptian murals fused with contemporary influences that produced a new representation of Egyptian identity (Abaza, 2012). For example, one painting by artist Alaa Awad, *The Funeral*, depicts a group of Ancient Egyptian women accompanying a sarcophagus which symbolises the mourning style of women in Upper Egypt. According to Awad, Egyptian women have always been expressive about mourning the dead, which involves certain customs such as tearing their clothes, extreme weeping, and smearing their bodies with mud (Abaza, 2012). Another iconic painting, in collaboration with artist Ammar Abu Bakr, is *The Hostesses*, which depicts women's empowerment and their role in the mass mobilisation against authoritarianism. Here, Awad paints seven cows behind the hostesses symbolising seven Gods of fortune, joy, femininity, goodness, and a brighter future. Abu Bakr painted three heads of army generals on a snake body as a form of an ongoing confrontation over freedom.

Artists played a vital role in visually documenting the violent confrontations with the army, projecting their works onto the walls as a means of preserving the urban memory associated with this significant space. One of Awad's paintings, for instance, depicts a chaotic clash of human figures together with a mixture of colourful identities and behaviours. These artworks created a unique and dynamic dialogue between the physical space, the art itself, and the people, collectively serving as a powerful reminder of the traumatic events that unfolded there. Furthermore, artists felt the desire to document the authorities' continuous efforts to erase the murals. Eventually, all murals on Mohamed Mahmoud Wall were erased by the authorities. The artworks can still be found online, captured through the lenses of professional photographers at the time. Despite not having lived as long as the Berlin Wall, the Mohamed Mahmoud Wall is, and will be, always remembered as the sight of the horrifying clashes that resulted in its identity as the landmark of civil resistance (Abdel-Rahman and Mahmoud, 2015).

A Reflection

Alaa Awad's artistic inspiration from ancient Egypt is fitting given Gilles Deleuze's exploration of ancient Egyptian art in *Francis Bacon: The Logic of Sensation*, particularly Chapter 14. Here, Deleuze highlights the haptic quality of ancient Egyptian art. Deleuze claims that these artworks are more akin to an "assemblage of bas-relief" (Deleuze, 1981: 122) than mere paintings. "Bas-relief" means that the relief is slightly projected from the background surface giving emphasis to its geometric essence, linking the hand and eye, and thus touch and vision. Interaction with such an artwork imposes a haptic function, signifying a deep, embodied engagement that is of a sensory nature, rather than operating on a more cognitive or representational level. Similarly, the graffiti seen in Cairo and Berlin, also exhibit this haptic quality by transforming public spaces of history and politics into sites that evoke collective memory and shared value. Social memory becomes tactile through the artwork on the walls and narrates transitions of culture over time—an "existential" challenge that includes the "construction of self-identity" (Halbwachs, 1992; Olick and Robbins, 1998; Jacobson, 2020: 5).

Graffiti was explored in this chapter to examine how visions of the cities of Cairo and Berlin become embedded and contested within the urban fabric, through a focus on walls in either city (Popczyk, 2015). In the case of the East Side Gallery, Berlin, the wall became a symbol of resistance. Through their artworks, the artists endeavoured to oppose the state, consumer culture, attempts to project luxurious and aspirational yet hollow images of life through architecture, and the wall itself, all whilst trying to evoke further resistance. There is, therefore, a complex network generated between people, space, and the artworks, what Tursić calls a "networked agency" (Tursić, 2019: 5). In Cairo, the Mohamed Mahmoud Wall worked to instil the culture of revolution, being emblazoned with images of hope, fear, and value. During the 18 days of the revolution, the interaction between people, the urban space, and the graffiti created a "people-space-artwork network", revealing fragments of Cairo's political and social realities. In both cases, graffiti played a pivotal role in producing aesthetic spaces that were not only artistic but deeply intertwined with cultural and political expressions.

Conclusion

Looking back to the scene described at the beginning of this chapter in *Wings of Desire*, Damiel's experience of his own existence symbolised an emotional response to a specific image that provoked cultural memory and identity. Damiel's sense of touch and taste in relation to the paintings of Thierry Noir on the Berlin Wall invited viewers to engage with the contextual memory evoked by the scene. With this in mind, this chapter has focused on graffiti and murals in both Berlin and Cairo as central expressions of collective memory and as mnemonic tools that embed these memories into the present.

In doing so, the chapter has added a sensory layer to understandings of murals and graffiti: the sense of touch or the "haptic", in Deleuzian terms. This idea of societies being touched by memories emphasises how art takes on a life of its own that emerges from the interaction between artists, their artworks, the city, and the public that comprise them. Focusing on the haptic and touch redefines urban aesthetics as a dynamic and sensory experience rather than a static and stable form. By considering urban aesthetics in this way, the urban form becomes a canvas for continuous storytelling—a place where history, culture, and personal narratives converge—that often have "the city" and how it should be understood, ordered, and lived out, as its stakes. The interplay between art, memory, and the environment creates a living and continuously evolving aesthetic, shaped by new interactions, or additional layers of meaning added by subsequent artists or viewers.

We suggest that thinking through the haptic and approaching aesthetics as a living and dynamic entity could provide architects and urban planners with a valuable lens for approaching and thinking about the aesthetics of memory around the city, particularly in contexts where history and identity are deeply intertwined with the physical environment. Indeed, the focus on graffiti has highlighted how public art serves as a living archive of societal narratives, blending past and present through its spatial and temporal dimensions. However, memory is not reducible to these artworks and vice versa, and it is for this reason that we have suggested that graffiti be understood in a mnemonic sense—in that it can work to prompt collective and individual memories. Architects and planners, who often work at the intersection of theory and practice, could draw from these insights to better understand the evolving "image" of a particular city. Integrating this framework into their work could encourage a design that resonates with local histories and cultural identities. What is more, such an approach might encourage professionals to engage with the physical and emotional connections people have with a city's past, translating these connections into meaningful elements within the urban aesthetic.

Notes

1 Though we acknowledge this reductive classification overlooks the full gamut of approaches, for instance, see the introduction to this edited volume.

2 Tahrir Square had its share of political demonstrations prior to 2011. In 1919, Egyptian mass protests rose against the British occupation, when the square was named Ismailia Square—having been named after Khedive Ismail in 1865. On 21 February 1946, protests took place at the square calling for the evacuation of British forces from the Nile Valley. The name "Tahrir" was officially used after Egypt became a republican nation during Nasser's era in the 1950s. Following the military defeat in the 1967 war, protests took place in the square calling for then-President Gamal Abdel Nasser to step down. Another significant protest was in January 1977 that resulted from raising prices, widely known at the time as the "Bread Riots". In March 2003, large protests took place in the square condemning the US invasion of Iraq. Ironically, around 20,000 protestors, who spent the whole night occupying the square, were violently dispersed by security forces the next day.

References

Abaza, M. (2012). 'Walls, Segregating Downtown Cairo and the Mohammed Mahmud Street Graffiti', *Theory, Culture & Society*, vol. 30, no. 1, pp. 122–139.

Abaza, M. (2013). 'Cairo Diary: Space-Wars, Public Visibility and the Transformation of Public Space in Post-Revolutionary Egypt'. In *Public Space, Media Space*, edited by Chris Berry, Janet Hardbord and Rachel O. Moore. Palgrave Macmillan: New York, pp. 88–110.

Abdel-Ghani, T. (2020). 'Film Intervention in Public Space: A Phylogenetic Spatial Change', *The Journal of Public Space*, vol. 5, no. 4, pp. 107–122.

Abdel-Rahman, A. and Mahmoud, R. (2015). 'The Post-revolutionary Effect on the Urban Harmony of Cairo's Built Environment in Relation to the Collective Memory of the Population: Urban Context of the "after revolution" between Contravention and Elaboration', *Ain Shams Engineering Journal*, vol. 7, pp. 1099–1106.

Alle, E. (2012). 'Spatial, Temporal and Social Dimensions of the Landscape influenced by Contemporary Art', *Mokslas – Lietuvos Ateitis*, vol. 4, no. 2, pp. 176–187.

Awad, S. (2017). 'Documenting a Contested Memory: Symbols in the Changing City Space of Cairo', *Culture & Psychology*, vol. 23, no. 2, pp. 234–254.

Bach, J. (2016). 'The Berlin Wall after the Berlin Wall: Site into Sight', *Memory Studies*, vol. 9, no. 1, pp. 48–62.

Bachelard, G. (1964). *The Poetics of Space*. Beacon Press: Boston.

Bayat, A. (2010). *Life as Politics: How Ordinary People Change the Middle East*. Stanford University Press: California.

Biggs, E. W. (2009). *Everyone Play: Sound, Public Space, and the (Re) making of Place*. A Thesis presented for the Doctor of Philosophy degree at the department of music at Princeton University, New Jersey.

Bobaroğlu, M. (2005). Eastern and Western Aesthetics. Available at: https://metinbobaroglu.net/en/eastern-and-western-aesthetics/ (accessed: 06 September 2022).

Bolt, B. (2010). *Art Beyond Representation: The Performative Power of the Image*. I.B. Tauris: London.

Bowen, T. (2013). 'Graffiti as Spatializing and Performance', *Rhizomes*, vol. 25, no. 1, pp. 1–15.

Boyd, C. P. (2017). *Non-Representational Geographies of Therapeutic Art Making*. Palgrave Macmillan: Melbourne.

Boyer, P. and Wertsch, J.V. (2009). *Memory in Mind and Culture*. Cambridge University Press: Cambridge.

Brooke, M. (2007). *"Mauerkunst, Lebenskunst": An Analysis of the Art on the Berlin Wall*. A Thesis presented for the Bachelor of Arts, Scripps College: Columbia.

Brosius, F. (2021). Art Against State Power – from the Street to Augmented Reality. Available at: https://www.art.salon/artworld/art-against-state-power-from-the-street-to-augmented-reality (accessed: 28 April 2023).
Burke, J. (2009). Berlin Wall Given a Facelift as Freedom Painters Return. Available at: https://www.theguardian.com/world/2009/apr/19/berlin-wall-freedom-painters (accessed: 23 August 2022).
Caldas, R. (2015). *Interpretation and Communication of Uncomfortable Heritage: the East Side Gallery*. A Thesis presented for the Master of Arts in World Heritage Studies, Brandenburg University of Technology, Cottbus.
Crinson, M. (2005). *Urban Memory: History and Amnesia in the Modern City*. Routledge: London.
Davies, D. (2017). 'Walls of Freedom: Street Art and Structural Violence in the Global City', *Rupkatha Journal on Interdisciplinary Studies in Humanities*, Vol. IX, no. 2, pp. 6–18.
Deleuze, G. (1981). *Francis Bacon: The Logic of Sensation*. Continuum: New York.
Deleuze, G. (1985). *Cinema 2: The Time-Image*. University of Minnesota Press: Minneapolis.
Ebeling, R. M. (2019). The History and Meaning of the Berlin Wall. Available at: https://www.aier.org/article/the-history-and-meaning-of-berlin-wall/ (accessed: 28 March 2023).
Foucault, M. (1971). *The Order of Things: An Archaeology of Human Sciences*. Pantheon Books: New York.
Fransberg, M., Myllylä, M., and Tolonen, J. (2021). 'Embodied Graffiti and Street Art Research', *Qualitative Research*, vol. 23, no. 2, pp. 1–18.
García-Doménech, S. (2015). 'Urban Aesthetics and Social Function of Actual Public Space: A Desirable Balance', *Theoretical and Empirical Researches in Urban Management*, vol. 10, no. 4, pp. 54–65.
Guattari, F. (1995). *Chaosmosis: An Ethico-aesthetic Paradigm*. Power Books: Sydney.
Halbwachs, M. (1992). *On Collective Memory*. University of Chicago Press: Chicago.
Hansen, S. and Flynn, D. (2015). 'Longitudinal Photo-documentation: Recording Living Walls', *Street Art & Urban Creativity Journal*, vol. 1, no. 1, pp. 26–31.
Harrison, P. (2000). 'Making Sense Embodiment and the Sensibilities of the Everyday', *Environment and Planning D: Society and Space*, vol. 18, no. 4, pp. 497–517.
Harvey, D. (1989). *The Condition of Postmodernity*. Blackwell: Oxford.
Harvey, D. (2012). *Rebel Cities: From the Right to the City to the Urban Revolution*. Verso: London.
Hoelscher, S. and Alderman, D. H. (2004). 'Memory and Place: Geographies of a Critical Relationship', *Social & Cultural Geography*, vol. 5, no. 3, pp. 347–355.
Jacobson, M. (2020). 'Graffiti, Aging and Subcultural Memory – A Struggle for Recognition through Podcast Narratives', *Societies*, vol. 10, no. 1, pp. 1–17.
Johannis, T. (2011). 'The Aesthetic Impact of Graffiti Art on Modern Greek Urban Landscape', *Ángulo Recto. Revista de estudios sobre la ciudad como espacio plural*, vol. 3, no. 2, pp. 17–35.
Jones, J. (2014). Thierry Noir: The First Graffiti Artist Fired up by the Berlin Wall. Available at: https://www.theguardian.com/artanddesign/jonathanjonesblog/2014/apr/03/thierry-noir-graffiti-artist-berlin-wall (accessed: 10 September 2022).
Koetter, F. (1980). 'Notes on the In-Between', *Harvard Architecture Review*, vol. 1, pp. 63–73
Latham, A. and McCormack, D. P. (2004). 'Moving Cities: Rethinking the Materialities of Urban Geographies', *Progress in Human Geography*, vol. 28, no. 6, pp. 701–724.
Latham, A. and McCormack, D. P. (2009). 'Thinking with Images in Non-representational Cities: Vignettes from Berlin', *Area*, vol. 41, no. 3, pp. 252–262.

Lillios, K.T. and Tsamis, V. (2010). "Introduction". In *Material Mnemonics: Everyday Memory in Prehistoric Europe*, edited by Katina T. Lillios and Vasileios Tsamis, pp. 1–10. Oxford: Oxbow Books.
Lobo, M. (2019). 'Encountering the City: Haptic Images of Suburban Darwin', *Area*, vol. 53, no. 2, pp. 201–210.
Lorimer, H. (2005). 'Cultural Geography: The Busyness of Being "more-than-representational"', *Progress in Human Geography*, vol. 29, no. 1, pp. 83–94.
Loughran, K., Fine, G.A. and Hunter, M.A. (2015). "Urban Spaces, City Cultures, and Collective Memories". In *Routledge International Handbook of Memory Studies*, edited by Anna Lisa Tota and Trever Hagen, pp. 193–204. Routledge: New York.
Massey, D. (2005). *For Space*. SAGE Publications: London.
Massumi, B. (2022). *Parables for the Virtual: Movement, Affect, Sensation*. Duke University Press: London.
McCormack, D. P. (2008). 'Geographies for Moving Bodies: Thinking, Dancing, Spaces', *Geography Compass*, vol. 2, no. 6, pp. 1822–1836.
McCormack, D. P. (2013). *Refrains for Moving Bodies: Experience and Experiment in Affective Spaces*. Duke University Press: Durham.
Mesch, C. (2008). *Modern Art at the Berlin Wall: Demarcating Culture in the Cold War Germanys*. Taurus Academic Studies: London.
Miller, I.L. (2002). *Aerosol Kingdom: Subway Painters of New York*. University Press of Mississippi: Mississippi.
Mitchell, D. (2003). *Right to the City: Social Justice and the Fight for Public Space*. Guilford: New York.
Nagati, O. and Stryker, B. (2013). *Archiving the City in Flux: Cairo's Shifting Landscape since the January 25th Revolution*. Cluster: Cairo.
Olick, J.K. and Robbins, J. (1998). 'Social Memory Studies: From "Collective Memory" to the Historical Sociology of Mnemonic Practices', *Annual Review of Sociology*, vol. 24, pp. 105–140.
Othman, S., Nishimura, Y., and Kubota, A. (2013). 'Memory Association in Place Making: A Review', *Procedia Social and Behavioural Sciences*, vol. 85, pp. 554–563.
Overdorf, J. (2016). Here's a Shocker: Repainting Part of the Berlin Wall Is a Sensitive Topic. Available at: https://theworld.org/stories/2013-11-11/heres-shocker-repainting-part-berlin-wall-sensitive-topic (accessed: 10 September 2022).
Petrov, N., Ezh, P., Piskunova, K. and Elagina, V. (2019). 'The Local Text of a City and Public Art: In Search of a Post-Soviet Identity', *Graffiti, Street Art & Urban Creativity Scientific Journal*, vol. 5, no. 1, pp. 28–42.
Pink, S. (2008). 'An Urban Tour: The Sensory Sociality of Ethnographic Place-Making', *Ethnography*, vol. 9, no. 2, pp. 175–196.
Pinto, S. (2017). 'The Wall Is Dead, Short Live Graffiti and Street Art! Graffiti, Street Art and the Berlin Wall's Heritage', *Graffiti, Street Art & Urban Creativity Scientific Journal*, vol. 3, no. 1, pp. 62–72.
Popczyk, M. (2015). 'The Aesthetics of the City-image', *Argument*, vol. 5, no. 2, pp. 373–386.
Postalcy, E., Ada, E., and Eren, O. (2006). 'The New Urban Memory', In *Proceedings of the 42nd International Society of City and Regional Planners Congress*, Istanbul, September 14–18.
Prior, R. (2019). As the Berlin Wall Was Falling, One Artist Concocted a Plan to Turn it into the World's Longest Outdoor Gallery. Available at: https://edition.cnn.com/style/article/east-side-gallery-berlin-trnd/index.html (accessed: 09 September 2022).
Pugh, E. (2015). 'Graffiti and the Critical Power of Urban Space: Gordon Matta-Clark's *Made in America* and Keith Haring's Berlin Wall Mural', *Space and Culture*, vol. 18, no. 4, pp. 421–435.

Rabat, N. (2011). 'Circling the Square: Architecture and Revolution in Cairo', *Artforum*, vol. 49, pp. 182–191.
Said, E.W. (2000). 'Invention, Memory, and Place', *Critical Inquiry*, vol. 26, no. 2, pp. 175–192.
Sanjuán, L.G. and Wheatley, D. W. (2010). "Natural Substances, Landscape Forms, Symbols and Funerary Monuments: Elements of Cultural Memory Among the Neolithic and Copper Age Societies of Southern Spain". In *Material Mnemonics: Everyday Memory in Prehistoric Europe*, edited by Katina T. Lillios and Vasileios Tsamis, pp. 10–40. Oxbow Books: Oxford.
Schmidt, L. and Klausmeier, A. (2004). *Wall Remnants – Wall Traces*. Westkreuz-Verlag: Berlin.
Street Art London, (2013). Interview: Thierry Noir. Available at: https://streetartlondon.co.uk/blog/2013/02/interview-thierry-noir/ (accessed: 28 March 2023).
Thrift, N. (2008). *Non-representational Theory: Space, Politics, Affect*. Routledge: Abingdon.
Tiwari, R. (2010). *Space – Body – Ritual: Performativity in the City*. Lexington Books: New York.
Totaro, D. (2002). 'Deleuzian Film Analysis: The Skin of the Film', *Off Screen*, vol. 6, no. 6. Available at: https://offscreen.com/view/skin_of_film (accessed: 28 March 2023).
Tursić, M. (2019). 'The City as an Aesthetic Space', *City*, vol. 23, no. 2, pp. 205–221.
Wang, F., Peng, Y., Wang, H.C., and Yin, F. (2016). 'Old Walls, Modern City: Research on Urban Memory of Disappearing Ancient Beijing City Walls', *International Journal of Tourism Cities*, vol. 2, no. 1, pp. 29–44.
Weizman, E. (2007). *Hollow Land: Israel's Architecture of Occupation*. Verso: London.
Wings of Desire. (1987). Directed by Wim Wenders. [Feature Film]. Argos Films: Paris, France.
Yang, J. (2019). 'A Sense of Japanese Aesthetics: The Role of Materiality in the Work of SANAA', In *2019 XJTLU International Conference: Architecture across Boundaries, KnE Social Sciences*, pp. 26–39.
Yoshitaka, O. (2013). 'What is 'the Haptic'? Consideration of *Logique de la Sensation* and Deleuze's Theory of Sensation', *Aesthetics*, no. 17, pp. 13–24.
Zimmerman, P.T. (2008). *Liminal Space in Architecture: Threshold and Transition. A Proposal for a Ballpark in Miami Beach, Florida*. A Thesis presented for the Master of Architecture Degree, The University of Tennessee, Knoxville.

8 Rancière, Aesthetics, and the Politics of the City-Scale*

Joe Blakey

Introduction

There are significant parallels between the questioning of the city's centrality in urban thought and the scale debate that emerged in human geography in the early 2000s. Both highlight concerns that these concepts—the city and scale—impose an epistemological framework that constrains our interpretation of phenomena. Moreover, the terms are also often equated in the "city-scale", as this chapter explores. This chapter argues that aesthetics can help overcome the impasse around the scale debate within human geography, and in doing so, provide insights for rethinking the city-scale in urban studies, while also addressing the critiques of city-centric thinking and the reification of scale.

Aesthetics means many things to urban scholars and geographers (Hawkins and Straughan, 2016), which I need not recount given this edited volume's introduction. Here, I draw upon Jacques Rancière's (1999) post-foundational approach to aesthetics to move past the impasse around scale. For Rancière aesthetics relates to "forms of perceiving the world and modes of relating to it" (Dikeç, 2012a: 274; see also Rancière, 2004: 304). In other words, Rancière (2009a: 56) suggests we are bound together by a "sensory fabric" shaping how we make sense of the world. Rancière seldom uses aesthetics explicitly, terming this sensory fabric as *le partage du sensible* (the distribution of the sensible), which politics, in turn, disrupts and transforms (Rancière, 1999). Developing Rancière's approach, I argue that scales, of which the city-scale can be part, serve as one such "common-sense" frame which shapes places and the roles played in society.

The "city-scale" is a term widely used within urban studies. For instance, Çağlar and Schille consider the city-scale to occupy "a relative position [...] on a field of power" reflecting "flows of political, cultural, and economic capital" and how these flows are shaped by local forces (Çağlar and Glick Schiller, 2010: 7). In their study of race-based readings of urban safety, Holman et al. (2023) argue that the city-scale serves as a crucial arena for implementing and

* This chapter has been adapted from an article originally published by Sage. See: Blakey, J., 2021. The politics of scale through Rancière. *Progress in Human Geography*, 45(4), pp.623–640. https://doi.org/10.1177/0309132520944487

DOI: 10.4324/9781003335993-8

opposing bordering practices. However, scholars are expressing concerns about city-centric thinking and "methodological cityism": scholarship that assumes "the city" as a self-evident unit of analysis (Brenner, 2019; Angelo and Wachsmuth, 2015). The city-scale then occupies an awkward conceptual space in urban studies: a scale so frequently invoked and assumed central to urban concerns, yet also something that perhaps distracts from urban phenomena as they actually occur. This chapter focuses on the question of scale more broadly, rather than the city-scale specifically, as the issues and arguments discussed are directly applicable to it. As the critiques around city-centric thinking have been covered extensively in the introduction to this edited volume (which is advisable to read first), this chapter will instead focus on the scale debate in geography, what Rancière's aesthetic approach offers, and the implications for the city-scale and city-centric thought specifically.

In particular, the chapter will consider what has been termed the "politics of scale" (Smith, 2010) to address the processes of constructing and contesting scales. Both "politics" and "scale" are entangled in extensive debates. Human Geography's scale debate, in the 1990s and early 2000s, entailed literature debating "the nature of scale": what it is, is not, and how it ought to be deployed (Brenner, 2001; Jones, 1998; Kaiser and Nikiforova, 2008: 538, Marston, 2000; Marston, Jones, and Woodward, 2005). By the mid-2000s, scale was increasingly seen as "chaotic" (Howitt, 2003), even prompting suggestions of its disciplinary expulsion (Marston et al., 2005). The debate arrived at an impasse around scale's relationship with ontology and the political consequences of our conceptions. Recently, debates surrounding "politics", "the political" and how they relate to ontology have also deepened (Barnett, 2017; Dikeç, 2005; Dikeç, 2012a; Featherstone and Korf, 2012; Meyer, Schetter, and Prinz, 2012; Swyngedouw, 2009; Swyngedouw, 2009; Swyngedouw, 2014). Herein, I suggest, lies an opportunity to develop some of this thinking to offer one way forward for the debates around scale and the city-scale by extension.

Whilst there can be no panacea for these divides, reading the "politics of scale" through the political and aesthetic thinking of Jacques Rancière can move us beyond the present theoretical impasse on scale's relative position to ontology, where scholars have expressed that in presuming scales precede social activity, they are reified, serving as an ontological "predetermination" (Marston et al., 2005: 422). Whilst for many Rancière's thought is useful in considering the "post-political" nature of contemporary governance (e.g. Swyngedouw, 2009), what I am drawn to here is his approach to political change more broadly (Davidson and Iveson, 2014), and in particular, his "aesthetic" solution to the problem of ontological reductionism (a concern many scale theorists and urban scholars share).

This aesthetic approach to politics allows us to consider how social activity is shaped by inherited, shared, "common-sense" whilst avoiding ontological theorising and narrowing political possibilities. Whilst, in such a reading, the city-scale—and scales more generally—do precede social activity, there is nothing "natural" or ontologically given about them and they can always be disrupted by politics. Moreover, to avoid epistemologically reifying scales and the city-scale, I argue that we should draw upon Rancière's method of equality and focus on those moments where they are contested.

This chapter proceeds as follows. Next, Human Geography's scale debate is overviewed, highlighting how thinkers understand scale in relation to ontology, with the city-scale reflecting these broader disputes. In the third section, I introduce Rancière's political thought and his approach to avoiding ontologising political thinking. I utilise this understanding to reconsider the ontological disagreements of the scale debate, offering his aesthetic solution as a way forward. In the fourth section, I consider the implications of Rancière's aesthetic approach for the politics of scale and the city-scale. I argue that by keeping scales and the opportunity for their contestation in parallax view, a Rancièrian approach can more fully account for scale as a "product and a progenitor" of social activity (MacKinnon, 2010: 22). Finally, I conclude by suggesting that whilst we do not encounter scales "'up there' in a vertical imaginary but on the ground, in practice", this is not to say that these shared, vertical imaginaries do not matter (Marston et al., 2005: 420). Rather, what is vital for urban scholars and scale theorists to consider is how common-sense imaginaries of the city-scale, and scales more generally, are met with dissensus "on the ground" (Rancière, 2016a: 159).

An Ontological Dispute: Human Geography's Scale Debate

Towards a "Politics" of Scale

Before the 1980s, "scale" was largely taken for granted as a way of establishing spatial hierarchy and seen as part of a spatiotemporal ordering system hardwired into the mind (Herod, 2011). However, materialist interventions by Peter Taylor (1982) and Smith (1982) ignited the debate. Since this debate is well-documented (c.f. Herod, 2011, Jonas, 2015, MacKinnon, 2010, Marston et al., 2005), I trace only the key contours here.

Smith is credited with "profoundly revolutioni[sing ...]" scale in the early 1980s (Jones et al., 2017: 138–139), identifying the urban, national, and global as scales where uneven development can be observed (see Smith, 2010: Ch.4, Ch.5). Whilst Smith does not explicitly focus on the city-scale, his work suggests that scales, including the city-scale, are socially produced through the workings of capital, are "nested rather than hierarchical" (Smith, 1992b: 66), and produced rather than given (Smith, 2008; Smith, 2010). This led Marxist political economy to investigate how scales are shaped, considering their varying significance over space and time (Sheppard and McMaster, 2008).

Smith recognised these early theorisations were somewhat "wooden" as they focused on political economy (Smith, 2011: 262), and so introduced and popularised the "politics of scale" to capture how actors attempt the social construction of scale (Smith, 2010; Brenner, 2001; Jones et al., 2017, Kaiser and Nikiforova, 2008). The politics of scale highlights the interplay of hegemonic power arrangements and political struggle, where scales not only constrain and exclude but can be "a weapon of expansion and inclusion" (Smith, 1992a: 61, 78). In his work on gentrification resistance in New York City, Smith describes how the homeless community of Tompkins Square Park faced a curfew, sparking solidarity squats across the Lower East Side under the banner of "Tompkins Square Everywhere". This was evidence, for Smith (2010), of how actors "jump scale to 'elevate themselves to the next scale up the hierarchy'" (Smith, 2010: 232) and illustrates the political component of the politics of scale (Newstead, Reid, and Sparke, 2005).

The politics of scale literature has examined "the nature of scale"—what scale is, is not, and how rescaling takes place (Kaiser and Nikiforova, 2008: 538). Scholars have explored the social, political, cultural, and economic processes through which scales are redefined (Brenner, 1998; MacKinnon, 2010; Smith, 1993; Swyngedouw, 1997). These processes include scale jumping; scale bending, where "entrenched assumptions about what kinds of social activities fit properly at which scales are [...] challenged and upset" (Smith, 2008: 193); and glocalisation, "the contested restructuring of [...] the national scale" both upwards and downwards (Swyngedouw, 2004: 37).

However, for Marston, the 1990s politics of scale literature had become overly "[p]reoccupied with questions of capitalist production [...and] therefore tells only part of a much more complex story" (Marston, 2000: 233). An example is Brenner's (1997) study of post-war planning policies in the Federal Republic of Germany (1960–1990), considering how scales were reshaped under changing configurations of state and capital. Marston (2000) highlights the oversight of patriarchal gender relations in the politics of scale, advocating for a more nuanced understanding that considers not only capitalist production but also social reproduction and consumption.

Scholars have debated the "vertical" and "horizontal" dimensions of scale. Brenner (1998) introduced "scale structuration" to highlight the links between scales and their (re)hierarchisation. Some find this vertical focus too rigid, favouring networked, horizontal, and relational understandings (cf Cox, 1998; Dicken, 2004). Others have combined horizontal and vertical approaches (Amin, 2002; Brenner, 1998; Leitner, 2004). However, perhaps the most central debate has concerned the ontological status of scale (Herod and Wright, 2008), to which I now turn.

Scale: Ontological, Epistemological, Nonsensical?

In 1998, Katherine Jones argued that scale is a "representational trope", framing what is knowable and the meaning of space (Jones, 1998). This raised the question of whether scale is epistemological—a way of thinking about the world—or whether it is ontological—a fundamental "given" of the world. Reflecting that scales "as a mode of understanding" did not exist prior to becoming a trope for making sense of space, Jones (1998: 28) argued that scales are epistemological. Assuming scales to be ontological, Jones (1998) contended, naturalises them and limits our ability to analyse how they frame space. As Marston (2000: 220) notes, interest in scale was gravitating toward constructionist frameworks and "the rejection of scale as an ontologically given category". Jones (1998) set the stage not only for debates around *how* we can avoid reifying scale in ontology, but *who* is doing so.

Inspired by Jones' (1998) epistemological understanding, scholars began conceptualising scale as a discursive or representational device (Ansell, 2008; Delaney and Leitner, 1997; Jones, 1998; Kurtz, 2003: 893; Mansfield, 2005; Moore, 2008), often using horizontal and practice-based understandings (Bulkeley, 2005; Collinge, 2006; Leitner, Pavlik, and Sheppard, 2002; Marston et al., 2005). Kurtz (2003), for instance, noted how scalar frames serve as strategic devices for representing controversies, with counter-frames challenging them. Such approaches rejected the idea of scales as preceding social activity, which would grant scales a problematic "ontological priority" (Herod, 2001: 66; Herod and Wright, 2008). Others, more sympathetic to "inherited scales", explored performativity approaches, analysing how scale functions as a tool of governmentality, producing "scale effects" while also exposing "gaps and fissures" that destabilise these scalar frameworks (Kaiser and Nikiforova, 2008).

However, even scale as an epistemological construct would also be challenged by Marston et al.'s (2005) *Human Geography Without Scale*. Observing vast divergences in definitions of scale, the authors declared the concept "chaotic" (Marston et al., 2005). Central to this claim was the supposed dichotomy between the dominant perspective of scale as a "nested hierarchy of differentially sized and bounded spaces" (Marston et al., 2005: 416–417) and those, inspired by Latour and Deleuze (Jones, Woodward, and Marston, 2007), who opposed this "vertical" understanding, advocating a horizontal networked approach (Jones, Woodward, and Marston, 2007). Marston et al. (2005: 422) argue that both employ stifling *a priori* scalar imaginaries that overlook socio-spatial processes, with horizontal approaches supplementing "one ontological-epistemological nexus (verticality) with another (horizontality)". They also reject attempts to hybridise a vertical and horizontal understanding (Brenner, 1998), arguing that the "foundational weaknesses" of vertical approaches cannot be fixed by "integrating them with network formulations" (Marston et al., 2005: 422, 417). This critique has profound implications for how the city-scale is understood. By assuming the city as a pre-given scale of analysis, scholars risk reifying it as a self-evident entity, rather than interrogating the urban processes occurring on the ground. In this regard, the concerns around the city-scale are emblematic of the broader concerns with scale.

Marston et al. (2005) conclude that a spatial ontology involving scale "limits entry points into politics [...] by pre-assigning it a restricted framework for resistance" (Marston et al., 2005: 427). Instead, they propose eliminating scale as an ontological "predetermination", advocating a "flat", site-based ontology more concerned with relations, practices, and processes (Jonas, 2006). They argue this allows us to account "for socio-spatiality as it occurs" without imposing predetermined categories like scale (Marston et al., 2005: 425). The call has received mixed responses. Some embraced it (cf. Collinge, 2006; Springer, 2014; Escobar, 2007), with Springer (2014: 408) agreeing that "scale represents a theoretical distraction", pulling focus away from the grounded particularities of everyday life in favour of abstraction.

Elsewhere, Leitner and Miller (2007) and Jonas (2006) contested that verticality need not imply top-down hierarchical power relations, and power could flow from the bottom-up or in both directions (Collinge, 1999; Sayre, 2005). Others, such as Leitner and Miller (2007), have argued that recognising existing scalar structures like the city-scale is vital to developing strategies of resistance. While agreeing that scale is not an ontological category of space (Jones et al., 2017), Smith contends that replacing contingent understandings of scale, which he views as organising social difference, with horizontal spatial conceptions abolishes power distinctions through "wishful thinking" (Smith, 2015: 964). For this reason, the complete removal of scale has been criticised for reinforcing unequal scalar power relations (Kaiser and Nikiforova, 2008).

Some attempts at reunification have emerged. Moore (2008: 214) suggests abandoning scale as an analytical category to avoid ontological theorising, whilst focusing on how scalar orders become solidified "in consciousness and practice". MacKinnon (2010), drawing from a critical realist perspective, argues that scales arise from both capitalist restructuring and social practice, attempting to find common ground. For MacKinnon, "the so-called politics of scale are not fundamentally 'of' (about) scale" (2010: 32) but about how scale is deployed towards various ends and so advocates "scalar politics" to capture this, emphasising the need to understand how actors try to "ontologically 'fix' or 'undo' scales as material expressions of emergent power relations" (MacKinnon, 2010: 30).

The question of how scale ought to relate to ontology is central to the scale debate. Compare Smith's insistence that in asserting that scales precede social activity, he is not "proposing some ontological system of scales" (Smith, 1992b: 66; Jones et al., 2017), with others charging this as ontological reification (c.f. Herod, 2011: 45–46; Herod and Wright, 2008: 10–11; Jones, 1998; Moore, 2008). Scholars also contend that one's stance on the ontological status of scale has political implications, despite scale theorists often referring to different things as "political" (Castree, Featherstone, and Herod, 2008). For instance, Marston et al. (2005) argue that a "flat" ontology is more politically "open", while Leitner and Miller (2007) argue that recognising pre-existing scales is crucial for politics. The subsequent section suggests one way to move beyond these ontological and political divides.

Reframing Scale with Rancière

Police, Politics, and Ontological Principles

This section illustrates how reading scale through Rancière's "post-foundational" and aesthetic political framework offers one way forward. I turn to Rancière specifically because of his approach to ontology in relation to politics, which I will outline briefly. I argue that this approach provides valuable insights for geographers and urban scholars aiming to avoid reified notions of both scale and city-scale, while still recognising their persistent role in our collective "common sense".

In *Disagreement: Politics and Philosophy*, Rancière (1999) argues that the traditional pursuit of philosophy has been to decide what is and is not ontological. Radically, Rancière proclaims he is "not a political philosopher" (Rancière, 2003: np) and abstains from theorising an "ontological principle of politics" (Rancière, 2009b: 119). Rancière (1999) sees political philosophy as contradictory where theorising what a political process *is* imposes rigid ontological boundaries that paradoxically constrain politics, supplanting "the anarchic disorder of politics with [...the] hierarchical order of the philosopher" (Chambers, 2011: 305). Political philosophy becomes "an attempt to suspend the destabilizing potential of the political [...by] fixing the rules of political competition" (Žižek, 2006: 72). This apprehension about politics mirrors both Marston et al.'s (2005): 427) apprehension that scale theorising limits "the openness of the political" and concerns that assuming the city-scale constrains our understanding of urban phenomena. However, Rancière is distinct in that he does not refute the value of understanding pre-existing yet contestable ways of managing and making sense of the world including, I will argue, scale and city-scale.

This twin appreciation for the lack of any ontological principle for how society ought to be ordered *and* the contingent ways in which it is has led Marchart (2007) to label Rancière a "post-foundational" political theorist. For Marchart (2007), post-foundational political thinkers can be distinguished through their rejection of "foundationalist" perspectives which claim that there is a "natural" or ontological principle for how society ought to be organised. Post-foundationalists can also be distinguished from anti-foundationalists who reject the existence of all "foundations", considering them a barrier to thinking and enacting emancipatory politics. Instead, post-foundational thinkers assert that whilst there is no ontological principle for ordering society, it is nonetheless valuable to consider socially constructed "foundations" as necessary, contingent, and contestable attempts to order society, space, and time (Arditi, 2019; Marchart, 2007; Sparke, 2005).

A diverse set of post-foundational political thinkers have gained traction amongst geographers and urban scholars (Blakey et al., 2022; Blakey, 2024; Landau-Donnelly and Pohl, 2023; Wilson and Swyngedouw, 2014). However, Rancière adopts a distinct political lexicon. Politics is a "deviation from [... the] normal order of things" (Rancière, 2001: 18) which, in turn, is what Rancière calls the police order. This is not to be confused with the police *force*. The police refer to the ordering of space and time as "the seemingly natural order of things" (Dikeç, 2005: 174). Drawing upon a Kantian understanding of aesthetics as that which relates to sense-perception (Kant, 2007 [1790]), Rancière argues that the police order has a corresponding aesthetic configuration which normalises certain ways of seeing, sensing, feeling, acting, speaking, and being in the world. He terms this aesthetic configuration as *le partage du sensible* (which translates as both the partition and sharing of the sensible) (Dikeç, 2012a). This distribution of the sensible refers to a shared:

> system of self-evident facts of sense perception that simultaneously discloses the existence of something in common and the delimitations that define the respective parts and positions within it [...] a distribution of spaces, times, and form of activity.
>
> (Rancière, 2004: 12)

The distribution of the sensible refers to *framings of* time and space that arise in particular historical and geographical contexts (Dikeç, 2013). The police, therefore, can be understood as an incumbent governance order that is symbolic with material manifestations (Dikeç, 2012b; Dikeç and Swyngedouw, 2017). As Dikeç (2013: 82) observes, two premises structure Rancière's political thinking: "radical contingency and radical equality". Politics is always a possibility precisely because there is no "natural", ideal, or ontologically given way to order society, even though the police order might *appear* all-inclusive or natural (see Rancière, 2000: 124). Politics occurs when the police order is disrupted, declared as "wrong", and a *re*distribution of the sensible occurs (Dikeç, 2005; Rancière, 1999). For this to occur, "a polemical common space" must be set up, where the established order of things (the police) is challenged by a different logic in the name of equality (Dikeç, 2005: 178). This manifests as a dissensus, "the presence of two worlds in one", between the police distribution of the sensible and the logic of politics (Rancière: 2001: np). Equality, here, is not an end to be strived for but is instead a presupposition: politics is a verification of equality (Dikeç, 2012a; Rancière, 1999).

Moving Beyond the Ontological Dichotomy

Jones et al. (2017) recall a 2005 Association of American Geographers (AAG) debate around Marston et al.'s (2005) *Human Geography Without Scale*. Smith responded to the claim that "scale is not an ontological category of space" (Jones et al., 2017: 147), declaring, "I agree with you, but you still have to understand that scale exists!", a statement that Jones et al. remark is "somewhat contradictor[y]" (2017: 148). Jones et al.'s (2017) comment is worth recounting as it exemplifies the ontological dichotomy at play in Marston et al.'s (2005) argument. The choice presented is between a "foundational" concept of scale as an "ontological given category" (Marston et al., 2005: 419, Jones et al., 2017: 148) or the proposed "anti-foundationalist" eradication of scale.

Under a Rancierian approach, the rejection of scale as an ontological category and the acceptance of "already-existing" scales need not be seen as mutually exclusive. Rancière's post-foundationalist approach enables us to consider the role of scale, not as an "ontological category of space" (Jones et al., 2017: 147), but as part of a socially constructed police order—a normalised way of relating to the world that is inherited and contestable (Dikeç, 2017). In his conception of the police, Rancière seeks not to essentialise any given social order, but merely to recognise them alongside their "lack of foundation" in ontology and "the sheer contingency of any social order" (Rancière, 1999: 16). Rancière's approach is post-foundational precisely because he recognises that commonly held, existing ideas which work to order society (such as the city-scale and scale generally) can be quite rigid, and as they lack an ontological principle, they can always be contested.

Rancière offers an unexplored angle for approaching "already-existing" scales, be they the city-scale or otherwise, moving beyond the ontological dichotomy presented by Marston et al. (2005). Asserting that scale is part of a police order shares much with epistemological approaches in pointing to scale as a socially constructed representational or discursive device asserting or contesting scale (Delaney and Leitner, 1997). However, MacKinnon (2010: 22) has argued that focusing on practice, and the discursive articulation of scales, provides a "one-sided view" of the contingent "product and [...] progenitor" relationship that I argue Rancière is able to better capture. MacKinnon (2010) has similarly argued that the charge of ontological reification against those asserting that scales precede social activity is overdrawn. As I elaborate in the subsequent section, Rancière offers a way to consider inherited ideas about scales without risking ontological reductionism.

However, Rancière's approach is no panacea as there are limitations to ontological debate (Leitner and Miller, 2007). Moore's (2008) comments provide a useful example to this end, suggesting that despite his appreciation for the social construction and contingency of scale, Smith cannot be absolved from ontological theorising. Moore (2008) suggests Smith misunderstands ontology, citing Elden (2005), for whom ontology is about "how it all hangs together" (Searle, 1995: xi). But such a charge is dependent on what one understands ontology to be. For Rancière, it is the lack of an ontological principle that necessitates the contingent and contestable socio-spatial orders that he calls the police. So, whilst adopting a Rancièrian approach is not a silver bullet, it offers an unexplored angle, enabling a consideration for contingent scalar imaginaries *and* their lack of an ontological principle.

An Aesthetic Approach to (City) Scales

Whilst Rancière (1999) shares Marston et al.'s (2005) concern for the political ramifications of ontological reductionism, his solution is distinct. Instead, "aesthetics takes on a leading role" whilst ontological theorising is avoided (Ieven, 2009: 60). Under Rancière's aesthetic approach, we do not apprehend the world in a "flat" manner and instead inherit a "common-sense" way of "perceiving and making sense of the world"—the police "distribution of the sensible" (Dikeç, 2012a: 277). But what difference does it make to insist scale is part of the distribution of the sensible rather than an ontological entity? As Moore has suggested:

> many terms [...] are deeply constitutive of the common-sense 'folk sociologies' [...] by which we make sense of the world. As such they tend to be reified in social thought as essential and natural entities, and these reified understandings are often uncritically adopted by social scientists as categories of analysis.
>
> (2008: 207)

Moore (2008: 208) asserts that "however contingent and fluid" we argue scales to be, "once socially constructed they are treated as every bit as real and fixed as ontological givens". But the masterstroke within Rancière's framework is that he keeps *common-sense* ways of making sense of the world (the police) and the potential for *different* ways to make sense of the world (politics) in parallax view. By extension, then, we might consider common-sense imaginaries around (city) scales and how they are contested hand-in-hand to avoid ontologising them. Rancière terms this approach as a "method of equality" where, in considering the equal capacity of everybody to bring about politics, he seeks out "egalitarian articulations of an alternative world within the world as it is" (Davidson and Iveson, 2014: 7). He is interested in dis-*sensus*, a "division inserted in 'common sense': a dispute over what is given and about the frame within which we see something as given" (Rancière, 2010: 77). Politics brings about a *redistribution* of the sensible (Tolia-Kelly, 2019), such that it "is aesthetic in principle" (Rancière, 1999: 58). As Derickson (2017) argues, Rancière's method of equality compels us to reject understandings of spaces as *intrinsically* political and compels us to instead consider the political processes through which spatial imaginaries (such as scale) become denaturalised or how this is resisted. Considering scales through Rancière's aesthetic approach means we can attend to how scales are socially constructed, how shared and pre-existing ideas about scales shape social practice, and their fallibility in face of politics. But what does this look like in practice?

Scales and Supplements

Rancière's explicit comments about scale are rare. However, in an interview with *BibliObs*, he is asked if "a people" has to be national in scale (Rancière, 2016b). He first responds that a "people, in the political sense, always constitutes itself as something apart from the state form of the people" (Rancière, 2016b: np). For Rancière (1999: 22–23), the "people" are those that are surplus to the population recognisable within the police and the possible subject of politics: those that declare a wrong and bring about a redistribution of the sensible. Instead, Rancière asserts that the national scale is "a collective symbolisation and, like any symbolisation, it is the stakes of a permanent struggle" by the people (Rancière, 2016b: np). The national scale does not "exist" in a metaphysical sense, but, as the space of the police and politics are "enmeshed", common-sense ideas of the national scale often become *the stakes* of political struggle (Dikeç, 2012b: 181).

As "the spaces of politics are necessarily enmeshed with the space of the police [...] there are a multiplicity of possible spaces at different scales" that can be politicised (Bassett, 2014: 888). Rancière's comments about the "world police" in the final pages of *Disagreement* substantiate this further, where he writes:

> The reign of globalization is not the reign of the universal [...] There is a world police [...] But there is no world politics. The 'world' can get bigger but the universal of politics does not get any bigger.
>
> (Rancière, 1999: 139)

For Rancière, the global scale is yet another "collective symbolisation" that is not universal or ontological because it can be infinitely transformed in acts of politics (see also Rancière, 2016a: 157–159). However, what is unclear here is if scales such as the national or the global—or in our case the city-scale—do form a "collective symbolisation", should one consider them *as* a police distribution of the sensible or as merely an ordering device *within* a distribution of the sensible? My sense is the latter.

Brenner's (2001) distinction between singular and plural applications of the politics of scale is useful to understanding why. In the singular usage, "the politics of scale denotes the production, reconfiguration or contestation of some aspect of sociospatial organisation *within* a relatively bounded geographical arena" (Brenner 2001: 599). Singular conceptions, therefore, observe how a particular scale is socio-historically constructed. In contrast, the plural usage of the term denotes the "production, reconfiguration or contestation of particular differentiations, orderings and hierarchies *among* geographical scales" (Brenner 2001: 600). In Agnew's (1997) discussion of political parties in Italy, for instance, he demonstrates how policing outlooks are intrinsically tied to their constructions of scale, such that "the boundaries they draw [...] contingent as they may be, define the geographical scales that channel and limit their poli[cing] horizons" (Agnew 1997: 118). One might simply equate this understanding with the distribution of the sensible and how it "demarcate[s ...] the spaces of the visible and the invisible, and articulates these allocations of parties and parts" (Rancière, 1999: 40). Such an understanding also implies a multiplicity of coexisting police distributions of the sensible. This is the understanding that Davidson and Iveson (2015: 552) advance in their own reading of "the city" through a Rancièrian frame, remarking that it is "a naturalized police order or 'distribution of the sensible'". However, this multiplicity is problematic if we dig deeper into Rancière's thinking.

Brenner has warned that this singular usage of scale should be used only "in question to its *embeddedness* or *positionality* within a broader scalar hierarchy" (Brenner, 2001: 600). Implicit in this singular usage of scale is, therefore, an appreciation for the "outside" of any given scale. But yet, the police "is a partition of the sensible whose principle is the absence of a void and of a supplement" (Rancière, 2001: 8). In other words, there is nothing that the police leaves unaccounted for, there can be no "outside" or remainder as everything is seemingly in its place (Rancière, 2010: 36). It is for this reason that I suggest that the city-scale, and indeed scales most generally, can only be conceived of *as part of* this broader aesthetic ordering. Moreover, this also implies that there cannot be a multiplicity of concurrent police distributions of the sensible. Consequently, I shall argue that city scales, and scales more generally, are *part* of this assignment of place by the police. Sevilla-Buitrago (2015: 10) has similarly remarked that the distribution of the sensible might entail "regimes of scale and place that determines and regulates our social being and forms of socialization".

Scales and Superimpositions

Let us return both to Rancière's comments and to an example to understand the implications of this approach. Rancière not only observes that scales, such as "the city" or "the national", are often the stakes of politics, but that "a people also constitutes itself locally, in relation to a given domination" (2016b: np). In *The Method of Equality*, for instance, Rancière writes that NGOs may "fight the powers of the world government" but that they do so "on the ground" (2019: 159). Rancière's point is that even though scales might be at stake, politics is always rooted in local "fractures" from which "egalitarian logic" works to disrupt the police (Rancière, 1999: 137).

Rancière gives the example of the Spanish *Indignados* who pitted themselves against the national-scale logic of austerity (Rancière, 2016b), a case well-rehearsed amongst Rancièrian thinkers in geography and urban studies (Bassett, 2014; García-Lamarca, 2017; Sevilla-Buitrago, 2015; Swyngedouw, 2018). The *Indignados* ("the outraged") are a self-titled group of anti-austerity activists who, following the arrest of 24 demonstrators during a protest against neoliberal economic reforms on 15 May 2011, set up camp in Puerta del Sol square in Madrid, Spain (Sevilla-Buitrago, 2015). As Sevilla-Buitrago (2015: 11) suggests, in setting up camp the movement reappropriated public space to create a "space of appearance", a space "from which a new democratising political sequence may unfold" (Swyngedouw, 2014: 174). This "space of appearance" broke with the given order of "where the voices of the people could be expressed, what they could say and how they could narrate and represent public space" (Sevilla-Buitrago, 2015: 11).

In demanding their right to equality, Puerta del Sol square became a "a polemical common space" (Dikeç, 2005: 178), a produced space from which politics may—or may not—arise (Davidson and Iveson, 2014; Dikeç, 2005, 2012a). This attempt to construct a "local" space of appearance for the *Indignados* also took place within the space of the police (Dikeç and Swyngedouw, 2017). A polemical space, therefore, is one where there are "two worlds in one and the same world" (Rancière, 2004: 304), a space where "two logics" for making sense of the world meet (Rancière, 1999: 32). On the one hand, there was a national-scale logic in Spain which identified that the cost of the crisis ought to be borne by the public (Bassett, 2014). On the other, the demonstrators provided "localised inscriptions" of a different logic, whereby the commodification of public space and the pandering to the financial markets is eradicated (Sevilla-Buitrago, 2015; Swyngedouw, 2014: 171). In this sense, Puerta del Sol square became a polemical space precisely because it became an aesthetic superimposition between this national-scale logic and the local scene, staged by the *Indignados*, which declares it as wrong (Rancière, 2016b).

But of course, this example is of the national scale. If we are to consider the city-scale in particular, we might similarly look to the seemingly privileged role of the city in work drawing upon Rancière and other post-foundational thinkers. Take, for instance, Dikeç and Swyngedouw's (2017): 7) claim that there is "an urgent need to rethink urban politics and urban political theory in ways that are much more sensitive to considering the city as an immanent site for nurturing political subjectivation, mediating political encounter, staging interruption and experimentally producing new forms of democratization". They further discuss how this might further involve the "nurturing of political imaginaries of what constitutes a good city" (Dikeç and Swyngedouw, 2017: 7). The authors do note that whilst conflicts are often "about" the city(-scale), they invariably take place in a context and set of processes that go well beyond it. Take, for instance, Karaliotas and Kapsali's (2021) work on Orfanotrofio's housing squat in Thessaloniki in the context of the wave of migration across Europe since 2015. The authors argue that the Orfanotrofio's housing squat in Thessaloniki provided both a physical home for the newcomers (a word deliberately chosen) and those squatting in solidarity with them, but also a political stage. The common-sense imaginary of the city-scale and its qualities, including what—or who—did and did not belong in it, was precisely the stakes of their struggle. As Karaliotas and Kapsali put it, Orfanotrofio "not only disrupted the exclusionary ordering of our cities by staging the equality of newcomers—migrants and refugees—and locals but also reconfigured the dominant partitioning of the urban sensible by putting a city of solidarity in confrontation with the city of exclusion" (Karaliotas and Kapsali, 2021: 400). Again, the case serves to highlight how two ways of making sense of the city-scale were placed in superimposition, between an exclusionary logic and one of equality.

In both cases, scale provides the spatialised language to consider how this polemical space emerged on the ground. Moreover, the risk of "already-existing" scales imposing a "cordoned register for resistance" (Marston et al., 2005: 427) is mitigated by actively considering such moments of dissensus, where common-sense surrounding scales is disrupted through an alternative logic. Both cases also highlight how individual scales—and therefore the city-scale—cannot be taken in isolation. Swyngedouw (2014: 171) remarks that Puerta del Sol square was an "embryonic" politicisation, inspiring similar movements across Spain. Reminiscent of the scale-jumping chants of "Tompkins Square Everywhere" (Smith, 2010), the *Indignados* claimed "[w]e're here, but anyway its global, and we're everywhere" (Swyngedouw, 2014: 182). Rancière similarly considers scale-jumping tactics during the civil unrest that spread through Paris in May 1968. Following the expulsion of the movement's leader, the German-Jewish Daniel Cohn-Bendit, from Paris in May 1968, Rancière notes how demonstrators "declared, against all police evidence, '[w]e are all German Jews'" (Rancière, 1999: 59). Be it in claiming that "we are all German Jews", that the *Indignados* are "global" or "Tompkins Square Everywhere", it is also clear that not only are ideas about individual scales often the stakes of political claims, but that claims are made against a wider "gestalt of scale" (Smith, 1987: 63) which is declared as "wrong" (Rancière, 1999: 6). Moreover, in doing so, the qualities of one scale (e.g. the polemical status of Puerta del Sol square) can be utilised to inspire claims to or around another (e.g. the Indignados being global). Similarly, in the case of Orfanotrofio, solidarity initiatives were set up across Greece and abroad (Karaliotas and Kapsali, 2021).

Finally, it is important to respond to the concern amongst commentators that Rancière's thought, and post-foundational political thinking more generally, leads to a diminished frame to consider the political (Beveridge and Koch, 2017). What does an approach which reserves the term politics for demands of equality mean for urban scholars thinking about the city scale and scale theorists? As Dikeç (2017) has responded to Beveridge and Koch's (2017: 32) claim that Rancière's "restricted" conceptualisation "reduces the realm of political action", one can instead argue the contrary—that this is an expansionary frame. In insisting that anyone can "become a political subject and anything a political issue", Rancière's frame is remarkably open, expanding our horizons for thinking about how spatial orders are reconfigured (Dikeç, 2017: 52). Equally, this does not stop us considering how common-sense surrounding the city-scale is transformed through already-existing governance mechanisms such as party politics. For this, I lean upon Marchart's (2011) term "minimal politics", as a reordering of power within the possibilities of what the police permits. Take, for instance, the minimal politics of the wave of devolution of political powers to cities that has swept across the UK since 2011. It is indeed important to hold on to a lexicon that can describe how changes within a police order occur, which nonetheless work to redefine our common-sense surrounding scales.

Conclusion: The City-Scale, Aesthetics, and a Method of Equality

This chapter has demonstrated how Rancière's aesthetic framework offers a novel way forward for Human Geography's scale debate, whilst also considering the implications for urban studies' concerns around city-centric thinking in relation to the city-scale. I have argued that divisions around scale often stem from how commentators position scale in relation to ontology. Take, for instance, those who consider scales as an "already partitioned geography" (Smith, 1993: 101) and those who reject this as ontological reification (MacKinnon, 2010; Moore, 2008). These ontological divides are perhaps nowhere more evident than in the debates that played out after Marston et al.'s (2005) proposed rejection of scale and proposal of a "flat" ontology in its place. This chapter has argued that Rancière's aesthetic approach, which he terms as the distribution of the sensible, enables us to move beyond this impasse and offers an alternative solution to this "problem of ontological reductionism" (Ieven, 2009: 60). Rather than treating scales as ontologically fixed, Rancière's framework positions scales like the city-scale as part of a common-sense ordering of space that is always open to political contestation.

I have argued that scales, including the city-scale, are part of what Rancière terms as the "distribution of the sensible"—an aesthetic order that encompasses the shared perceptions that shape our understanding of space and society. By adopting this lens, we can see how scales can be part of a common-sense ordering of society that is always open to disagreement on the ground. In this vein, our common-sense ideas about scale precede and shape social activity, working to frame time and space in different ways, along with what is seen as in and out of place. Politics is always a possibility because our common-sense around scales lacks an ontological principle. Rather than limiting "entry points into politics", scale as a "vertical imaginary" can form the *stakes* of politics (Marston et al., 2005: 427; 420). It is vital, therefore, to recognise existing, commonly held ideas about scale and the city-scale to understand how they are supplanted through acts of politics (Leitner and Miller, 2007).

Whilst Rancièrian thinkers within geography and urban studies have (often implicitly) utilised scale and the city-scale in divergent ways, in this piece I have suggested that it is important to consider any given scale not in isolation, but in relation to a wider distribution of the sensible. In keeping scales and the opportunity for their contestation in parallax view, a Rancièrian approach can more fully account for scale as a "product and a progenitor" of social activity than epistemological approaches (MacKinnon, 2010:22). Moreover, given that scales take the form of contingent, shared understandings that are at once symbolic with material manifestations (Dikeç, 2012b; Dikeç and Swyngedouw, 2017), this approach is able to bring together idealist approaches which understand scales as "mental contrivances" and materialist understandings of scales as produced through political struggle (Herod, 2003: 21). Nonetheless, as Leitner and Miller (2007) suggest, there are limitations to ontological debate and this approach cannot be considered a panacea.

What, then, are the implications for thinking about the city-scale, and what are the limits of such an argument? This chapter is certainly not arguing that the concern for the urban can be collapsed into some entity called "the city" or the "city-scale", nor is it arguing that the "city-scale" is inevitable or reducible to a specific set of qualities. Rather, it is to argue that the city-scale can nonetheless serve as a popular imaginary that can *seem* relatively foundational to everyday urban life, shaping its material and lived conditions. In other words, we should not attempt to distil the city-scale as something concrete and absolute, as for instance, Ernest Burgess did in his famous 1921 concentric rings diagram, that stretches from a centre to suburbia. Instead, building on Rancière, what I advocate here is seeing such visions as powerful yet contestable social imaginaries. This common-sense imaginary is not a portrayal of urban processes as such, nor is it inevitable, but it can wield a particular social power, in shaping urban processes, and who or what does or does not belong within it. Crucially, given that the city-scale has no ontological basis, it can always be transformed in acts of politics.

It is also important not to neglect the intent of both arguments, around scale and city-centric thinking. Those warning against scale-theorising as "ontological reification" and those concerned that the city-scale is too often presumed have often done so out of a concern that frames of analysis foreclose our ability to consider phenomena as it actually occurs and the opportunity to consider alternative configurations. This is indeed important to consider, and Rancière's approach can mediate these risks. To respond to these concerns, I wish to lean on a point raised by Davidson and Iveson (2015), who have also mounted a conceptual defence of the city most generally (rather than the city-scale) on the basis of Rancière. As Davidson and Iveson (2015) have argued, the full implications of Rancière's "method of equality" have not been attended to in Human Geography. His method focuses attention on those moments when our existing ideas about scales are disrupted through politics (Iveson, 2015).

The point here is that we ought not to solely consider common-sense on what scales "exist", but also those moments when scales are contested (Davidson and Iveson, 2015). Indeed, whilst as Marston et al. (2005: 420) rightly point out, we do not encounter scales "'up there' in a vertical imaginary, but on the ground, in practice", this is not to say that these shared, vertical imaginaries do not matter. Under the Rancièrian method, the task for scale theorists is to consider those moments when common-sense on these vertical imaginaries such as the city-scale are met with dissensus "on the ground" (Rancière 2016a: 159). Doing so allows scholars to avoid presuming or ontologically reifying existing imaginaries by focusing on those moments where there is the presence of "two worlds in one" (Rancière, 2004: 304), thus rendering both thought and the police order open to change. It is also important to remember that the "on the ground" politics around the city-scale might not solely occur in the space we conventionally call the city, but perhaps elsewhere, such as virtual spaces.

Finally, this chapter poses broader challenges. How, for instance, might other post-foundational political thinkers or aesthetic thinkers infuse understandings of the city-scale? Moreover, the scale debate's impact extends beyond the concept itself, with Marston et al.'s (2005) "flat", site-based ontology being part of a broader disciplinary shift towards relational approaches (Anderson et al., 2012; Jones, 2009). But as Gerhardt (2020) has warned, a focus on sites and relations alone may lead to the oversight of those forces that order them.

Rancière's aesthetic and dissensual approach, then, serves as a means to consider how common-sense imaginaries around the scale and city-scale are made meaningful whilst avoiding the risks of ontological reductionism and remaining sensitive to emergent politics around how things can always be otherwise (Jessop, Brenner, and Jones, 2008; Leitner, Sheppard, and Sziarto, 2008; Sevilla-Buitrago, 2015).

References

Agnew, J. (1997). The dramaturgy of horizons: geographical scale in the 'reconstruction of Italy' by the new Italian political parties, 1992–95. *Political Geography*, 16(2), pp. 99–121.

Amin, A., (2002). Spatialities of globalisation. *Environment and Planning A: Economy and Space*, 34(3), pp. 385–399.

Anderson, B., Kearnes, M., McFarlane, C. and Swanton, D., (2012). On assemblages and geography. *Dialogues in Human Geography*, 2(2), pp. 171–189.

Angelo, H. and Wachsmuth, D., (2015). Urbanizing urban political ecology: A critique of methodological cityism. *International Journal of Urban and Regional Research*, 39(1), pp. 16–27.

Ansell, N. (2008). Childhood and the politics of scale: Descaling children's geographies? *Progress in Human Geography*, 33(2), pp. 190–209.

Arditi, B. (2019). Fidelity to Disagreement: Jacques Rancière's Politics without Ontology. In *Distributions of the Sensible: Rancière, Between Aesthetics and Politics*, edited by S. Durham and D. Gaonkar, Evanston: Northwestern University Press, pp. 53–78.

Barnett, C. (2017). *The Priority of Injustice: Locating Democracy in Critical Theory*. Athens: The University of Georgia Press.

Bassett, K. (2014). Rancière, politics, and the occupy movement. *Environment and Planning D: Society and Space*, 32(5), pp. 886–901.

Beveridge, R. and Koch, P. (2017). The post-political trap? Reflections on politics, agency and the city. *Urban Studies*, 54(1), pp. 31–43.

Blakey, J., (2024). What repoliticisation means and requires: Creating the climate for disagreement. *Political Geography*, 115, p. 103222.

Blakey, J., Machen, R., Ruez, D. and García, P.M., (2022). Intervention: Engaging post-foundational political theory requires an 'enmeshed' approach. *Political Geography*, 99, p. 102689.

Brenner, N. (1997). State territorial restructuring and the production of spatial scale: Urban and regional planning in the Federal Republic of Germany, 1960–1990. *Political Geography*, 16(4), pp. 273–306.

Brenner, N. (1998). Between fixity and motion: Accumulation, territorial organization and the historical geography of spatial scales. *Environment and Planning D: Society and Space*, 16(4), pp. 459–481.

Brenner, N. (2001). The limits to scale? Methodological reflections on scalar structuration. *Progress in Human Geography*, 25(4), pp. 591–614.

Brenner, N., (2019). *New Urban Spaces: Urban Theory and the Scale Question*. Oxford University Press.

Bulkeley, H. (2005). Reconfiguring environmental governance: Towards a politics of scales and networks. *Political Geography*, 24(8), pp. 875–902.

Çağlar, A. and Glick Schiller, N. (2010). Introduction: Migrants and Cities In *Locating Migration: Rescaling Cities and Migrants*, edited by N. Glick Schiller and A. Çağlar. Cornell University Press, Ch1.

Castree, N., Featherstone, D., and Herod, A. (2008). Contrapuntal Geographies: The Politics of Organizing Across Sociospatial Difference. In *The SAGE Handbook of Political Geography*, edited by K. Cox, M. Low and J. Robinson. London: SAGE Publications Ltd, pp. 305–321.

Chambers, S.A. (2011). Jacques Rancière and the problem of pure politics. *European Journal of Political Theory*, 10(3), pp. 303–326.

Collinge, C. (1999). Self-organisation of society by scale: A spatial reworking of regulation theory. *Environment and Planning D: Society and Space*, 17(5), pp. 557–574.

Collinge, C. (2006). Flat ontology and the deconstruction of scale: A response to Marston, Jones and Woodward. *Transactions of the Institute of British Geographers*, 31(2), pp. 244–251.

Cox, K.R. (1998). Spaces of dependence, spaces of engagement and the politics of scale, or: Looking for local politics. *Political Geography*, 17(1), pp. 1–23.

Davidson, M. and Iveson, K. (2014). Occupations, mediations, subjectifications: Fabricating politics. *Space and Polity*, 18(2), pp. 137–152.

Davidson, M. and Iveson, K. (2015). Recovering the politics of the city: From the 'post-political city' to a 'method of equality' for critical urban geography. *Progress in Human Geography*, 39(5), pp. 543–559.

Delaney, D. and Leitner, H. (1997). The political construction of scale. *Political Geography*, 16(2), pp. 93–97.

Derickson, K.D. (2017). Taking account of the 'part of those that have no part'. *Urban Studies*, 54(1), pp. 44–48.

Dicken, P. (2004). Geographers and 'globalization': (yet) another missed boat? *Transactions of the Institute of British Geographers*, 29(1), pp. 5–26.

Dikeç, M. (2005). Space, politics and the political. *Environment and Planning D: Society and Space*, 23(2), pp. 171–188.

Dikeç, M. (2012a). Politics is sublime. *Environment and Planning D: Society and Space*, 30(2), pp. 262–279.

Dikeç, M. (2012b). Space as a mode of political thinking. *Geoforum*, 43(4), pp. 669–676.

Dikeç, M. (2013). Beginners and equals: Political subjectivity in Arendt and Rancière. *Transactions of the Institute of British Geographers*, 38(1), pp. 78–90.

Dikeç, M. (2017). Disruptive politics. *Urban Studies*, 54(1), pp. 49–54.

Dikeç, M. and Swyngedouw, E. (2017). Theorizing the politicizing city. *International Journal of Urban and Regional Research*, 41(1), pp. 1–18.

Elden, S. (2005). Missing the point: Globalization, deterritorialization and the space of the world. *Transactions of the Institute of British Geographers*, 30(1), pp. 8–19.

Escobar, A. (2007). The 'ontological turn' in social theory. A commentary on 'Human geography without scale', by Sallie Marston, John Paul Jones II and Keith Woodward. *Transactions of the Institute of British Geographers*, 32(1), pp. 106–111.

Featherstone, D. and Korf, B. (2012). Introduction: Space, contestation and the political *Geoforum*, 43(4), pp. 663–668.

García-Lamarca, M. (2017). From occupying plazas to recuperating housing: Insurgent practices in Spain. *International Journal of Urban and Regional Research*, 41(1), pp. 37–53.

Gerhardt, H. (2020). Engaging the non-flat world: Anarchism and the promise of a post-capitalist collaborative commons. *Antipode*, 52(3), pp. 681–701.

Hawkins, H. and Straughan, E. (2016). Introduction: For geographical aesthetics. In *Geographical Aesthetics*, edited by H. Hawkins and E. Straughan. London: Routledge. pp 1–17.

Herod, A. (2001). *Labor Geographies: Workers and the Landscapes of Capitalism*. New York: Guilford Press.

Herod, A. (2003). Scale: The local and the global. In *Key Concepts in Geography*, edited by S. Holloway, S. Rice, G. Valentine and N. Clifford. 2nd ed. London: Sage, pp. 217–235.
Herod, A. (2011). *Scale*. London: Routledge.
Herod, A. and Wright, M. (2008). Placing Scale: An Introduction. In *Geographies of Power*, edited by Herod, A. and Wright, M., Oxford: Blackwell, pp. 1–14.
Holman, N., Mace, A., Zorloni, D., Navarrete-Hernandez, P., Karlsson, J., and Pani, E. (2023). Race-based readings of safety in public space in Milan, the challenge for urban design. *European Urban and Regional Studies*, 30(3), pp.282–296.
Howitt, R. (2003). Scale. In *A Companion to Political Geography*, edited by J. Agnew, K. Mitchell, and G. O'Tuathail. Oxford: Blackwell, pp. 138–157.
Ieven, B. (2009). Heteroreductives–Rancière's disagreement with ontology. *Parallax*, *15*(3), 50–62.
Iveson, K. (2015). Building a city for "The People": The politics of alliance-building in the Sydney Green Ban Movement. *Antipode*, 46(4), 992–1013.
Jessop, R. Brenner, N. and Jones, M. (2008). Theorizing sociospatial relations. *Environment and Planning D: Society and Space*, 26, pp. 389–401.
Jonas, A.E.G. (2006). Pro scale: Further reflections on the 'scale debate' in human geography. *Transactions of the Institute of British Geographers*, 31(3), pp. 399–406.
Jonas, A.E.G. (2015). Scale. In *A companion to Political Geography*, edited by J. Agnew, K. Mitchell, and G. Toal. New Jersey: John Wiley & Sons. 138–157.
Jones, J.P. Leitner, H. Marston, S.A. and Sheppard, E. (2017). Neil Smith's scale. *Antipode*, 49, 138–152.
Jones, J.P. Woodward, K. and Marston, S.A. (2007). Situating flatness. *Transactions of the Institute of British Geographers*, 32(2), pp. 264–276.
Jones, K.T. (1998). Scale as epistemology. *Political Geography*, 17(I), pp. 25–28.
Jones, M. (2009). Phase space: Geography, relational thinking, and beyond. *Progress in human geography*, 33(4), pp.487–506.
Kaiser, R.J. and Nikiforova, E. (2008). The performativity of scale: The social construction of scale effects in Narva, Estonia. *Environment and Planning D: Society and Space*, 26(3), pp. 537–562.
Kant, I. (2007[1790]). *Critique of Judgement*. Translated by J.C. Meredith. Oxford: Oxford University Press.
Karaliotas, L. and Kapsali, M. (2021). Equals in solidarity: Orfanotrofio's housing squat as a site for political subjectification across differences amid the "Greek crisis". *Antipode*, 53(2), pp. 399–421.
Kurtz, H.E. (2003). Scale frames and counter-scale frames: Constructing the problem of environmental injustice. *Political Geography*, 22(8), pp. 887–916.
Landau-Donnelly, F. and Pohl, L. (2023). Towards a post-foundational geography: Spaces of negativity, contingency, and antagonism. *Progress in Human Geography*, 47(4), pp. 481–499.
Leitner, H. (2004). The Politics of Scale and Networks of Spatial Connectivity. In *Scale and Geographic Inquiry: Nature, Society, and Method*, edited by E. Sheppard and R.B. McMaster. London: John Wiley & Sons. Chapter 11.
Leitner, H. and Miller, B. (2007). Scale and the limitations of ontological debate: A commentary on Marston, Jones and Woodward. *Transactions of the Institute of British Geographers*, 32(1), pp. 116–125.
Leitner, H. Pavlik, C. and Sheppard, E. (2002) Networks, Governance and the Politics of Scale: Inter-urban Networks and the European Union. In *Power, Politics and Geography*, edited by A. Herod and M. Wright. Oxford: Blackwell.
Leitner, H. Sheppard, E. and Sziarto, K.M. (2008) The spatialities of contentious politics. *Transactions of the Institute of British Geographers*, 33, pp. 157–172.

MacKinnon, D. (2010). Reconstructing scale: Towards a new scalar politics. *Progress in Human Geography*, 35(1), pp. 21–36.

Mansfield, B. (2005). Beyond rescaling: Reintegrating the 'national' as a dimension of scalar relations. *Progress in Human Geography*, 29(4), pp. 458–473.

Marchart, O. (2007). *Post-foundational Political Thought: Political Difference in Nancy, Lefort, Badiou and Laclau*. Edinburgh: Edinburgh University Press.

Marchart, O. (2011). The Second Return of the Political: Democracy and the Syllogism of Equality. In *Reading Rancière*, edited by P. Bowman and R. Stamp, London: Continuum. 129–147.

Marston, S.A. (2000). The social construction of scale. *Progress in Human Geography*, 24(2), pp. 219–242.

Marston, S.A. Jones, J.P. and Woodward, K. (2005). Human geography without scale. *Transactions of the Institute of British Geographers*, 30(4), pp. 416–432.

Meyer, R. Schetter, C. and Prinz, J. (2012). Spatial contestation?–The theological foundations of Carl Schmitt's spatial thought. *Geoforum*, 43(4), pp. 687–696.

Moore, A. (2008). Rethinking scale as a geographical category: From analysis to practice. *Progress in Human Geography*, 32(2), pp. 203–225.

Newstead, C. Reid, C. and Sparke, M. (2005). The cultural geography of scale. In *Handbook of Cultural Geography*, edited by K. Anderson, M. Domosh, S. Pile, and N. Thrift. London: SAGE, pp. 485–497.

Rancière, J. (1999). *Dis-agreement: Politics and Philosophy*. London: University of Minnesota Press.

Rancière, J. (2000). Dissenting words. A conversation with Jacques Rancière (with Davide Panagia). *Diacritics*, 30, 113–126.

Rancière, J. (2001). Ten theses on politics. *Theory & Event*, 5(3), https://doi.org/10.1353/tae.2001.0028

Rancière, J. (2003). Comment and responses. *Theory & Event*, 6(4).

Rancière, J. (2004). The politics of literature. *SubStance*, 33(103), pp. 10–24.

Rancière, J. (2009a). Afterword. In *Jaques Rancière: History, Politics, Aesthetics*, edited by G. Rockhill and P. Watts. Durham, NC: Duke University Press, pp. 273–288.

Rancière, J. (2009b). A few remarks on the method of Jacques Rancière. *Paralax*, 15(3), 114–123.

Rancière, J. (2010). *Dissensus: On Politics and Aesthetics*. London: Bloomsbury.

Rancière, J. (2016a). *The Method of Equality*. Cambridge: Polity.

Rancière, J. (2016b). Getting Beyond Hatred: An Interview with Jacques Rancière. Available from: https://www.versobooks.com/blogs/2505-getting-beyond-hatred-an-interview-with-jacques-ranciere [Accessed 11th June 2020].

Sayre, N.F. (2005). Ecological and geographical scale: Parallels and potential for integration. *Progress in Human Geography*, 29(3), pp. 276–290.

Searle, J. (1995). *The Construction of Social Reality*. New York: Free Press.

Sevilla-Buitrago, A. (2015). Outraged spatialities: The production of public space in the #spanishrevolution. *ACME: An International E-Journal for Critical Geographies*, 14(1), 90–103.

Sheppard, E. and McMaster, R.B. (2008). *Scale and Geographic Inquiry: Nature, Society, and Method*. London: John Wiley and Sons.

Smith, N. (1982). Gentrification and uneven development. *Economic Geography*, 58(2), 139–155.

Smith, N. (1987). Dangers of the empirical turn: Some comments on the curs initiative. *Antipode*, 19(1), pp. 59–68.

Smith, N. (1992a). Geography, Difference and the Politics of Scale. In *Postmodernism and the Social Sciences*, edited by J. Doherty, E. Graham, and M. Malek. London: Palgrave Macmillan. 57–79.

Smith, N. (1992b). Politics: Contours of a spatialized homeless of geographical vehicles and the production scale. *Social Text*, 33(33), pp. 54–81.
Smith, N. (1993). Homeless/global: Scaling places. In *Mapping the Futures*, edited by J. Bird, B. Curtis, T. Putnam, and L. Tickner. London: Taylor & Francis, pp. 87–119.
Smith, N. (2008). Scale Bending and the Fate of the National. In *Scale and Geographic Inquiry*, edited by R.B. McMaster, and E. Sheppard. London: John Wiley and Sons, pp. 192.
Smith, N. (2010[1990]). *Uneven Development: Nature, Capital, and the Production of Space*. Georgia: University of Georgia Press. Third Edition.
Smith, N. (2011). Uneven development redux. *New Political Economy*, 16(2), pp. 261–265.
Smith, N. (2015). The future is radically open. *Acme*, 14(3), pp. 954–964.
Sparke, M. (2005). *In the Space of Theory: Postfoundational Geographies of the Nation-state*. University of Minnesota Press.
Springer, S. (2014). Human geography without hierarchy. *Progress in Human Geography*, 38(3), pp. 402–419.
Swyngedouw, E. (1997). Neither global nor local: 'glocalization' and the politics of scale. In *Spaces of Globalization. Reasserting the Power of the Local*, edited by K.R. Cox, pp. 137–166. London/New York: Guilford Press.
Swyngedouw, E. (2004). Globalisation or 'glocalisation'? Networks, territories and rescaling. *Cambridge Review of International Affairs*, 17(1), pp. 25–48.
Swyngedouw, E. (2009). The antinomies of the postpolitical city: In search of a democratic politics of environmental production. *International Journal of Urban and Regional Research*, 33(3), pp. 601–620.
Swyngedouw, E. (2014). Insurgent architects, radical cities and the promise of the political. In *The Post-political and its Discontents: Spaces of Depoliticization, Specters of Radical Politics*, edited by E. Swyngedouw and J. Wilson. Edinburgh: Edinburgh University Press, pp. 169–188.
Swyngedouw, E. (2018). *Promises of the Political: Insurgent Cities in a Post-political Environment*. Massachusetts: MIT Press.
Taylor, P.J. (1982). A materialistic framework for political geography. *Transactions of the Institute of British Geographers*, 7(1), pp. 15–34.
Tolia-Kelly, D. (2019). Rancière and the re-distribution of the sensible: The artist Rosanna Raymond, dissensus and postcolonial sensibilities within the spaces of the museum. *Progress in Human Geography*, 43(1), pp. 123–140.
Wilson, J. and Swyngedouw, E. (2014). Seeds of Dystopia. In *The Post-Political and its Discontents*, edited by J. Wilson and E. Swyngedouw. Edinburgh: Edinburgh University Press.
Žižek, S. (2006). Afterword. In *The Politics of Aesthetics*, edited by G. Rockhill. London: Continuum, pp. 67–68.

9 Cities and the Sixth Sense

The Aesthetics of Experimental Urbanisms

Julian Brigstocke

Introduction

The urban has long been a key site and frame for theorizing the politics of aesthetics, especially in the critical theory tradition of thinkers such as Simmel, Benjamin, and Kracauer. Approaching cities from the perspective of aesthetics foregrounds the sensory and embodied aspects of urban existence. Yet aesthetics remains an undervalued category in urban studies and geography, where "too often … our engagements with aesthetics involve impatient and arguably reductive appeals to their more 'serious' socio-economic and political dimensions" (Kingsbury, 2010: 56). This is not to deny that aesthetics is closely linked to politics and economy; Gassner (2019: 7) is right to suggest that "making sense of the world and acting in the world require aesthetic forms; that aesthetic forms and aesthetic judgements introduce something new into the world and that they are, therefore, linked to radical politics". In this chapter, I explore an Arendtian approach to aesthetics that focuses on the imagination, the everyday, the provisional, the processual, the performative, and the micropolitical, rather than as consolidated political forms. Doing so allows an aesthetics of the city to refuse the lure of aesthetic reductions of the city to spectacularized, static images, monuments, landscapes, and political forms, and to emphasize what an aesthetics of the city adds above and beyond a sensory analysis of the city.

Much work on urban aesthetics draws attention to the sights, sounds, smells, and tastes of the city, as well as how the city touches and is touched by the body (Cowan and Steward, 2007; Diaconu et al., 2011; Goonewardena, 2005; Law, 2001). Studies of geographical aesthetics also go beyond the five senses to consider the "somatic" senses such as proprioception, kinaesthesia, and the vestibular sense (Paterson, 2007). A long tradition of urban thought also connects the city to a kind of sensory overload (Mubi Brighenti and Pavoni, 2019), where an onslaught of sensory, affective, and cognitive demands ("hyper-aesthesia") results in a protective anaesthesia, supposedly damaging urban dwellers' capacities for self-awareness, individualization, and care for others (Buck-Morss, 1992; Hayles, 2007; Stiegler, 2010). Others suggest a more cautious approach, pointing to the need for finer-grained empirical analyses of the connections between diverse aesthetic and affective states and urban environments (Amin and Richaud, 2020).

DOI: 10.4324/9781003335993-9

In this chapter, I contribute to an analysis of aesthetics and the city by exploring Hannah Arendt's notion of a "sixth sense" – a sense that is not connected to a sensory organ, and does not itself generate sensation, but instead creates a distinctive feeling of "realness". In Arendt's thought, the essence of the sixth sense, or "community" sense, is to be able to think and feel the world from the standpoint of others. The sixth sense is not tied to sensory experience, and does not unite the senses as a kind of "common sense". Rather, it is tied to imagination: the capacity to occupy the perspectives of others. It is this imaginative travelling amongst others that makes the world feel dimensional, real, something shared in common with others. According to Arendt, this sixth sense is mysterious but uniquely important. We have neither awareness nor memory of its emergence, and yet it is the sense that uniquely fits us for being in the world. As a *common* faculty – one which humans use as part of their shared, plural existence – the sixth sense is always a little mysterious and opaque (Burks, 2002). Yet, I will argue in this chapter that it is this sixth sense, the sense for reality, that underlies a politics of aesthetics capable, through the exercise of imagination, of nurturing exposure to plurality, diversity, and the performance of freedom. Conversely, as Arendt and Kroh (1964) make clear in their account of the "banality of evil", the dulling of this sense – the decline of an imaginative capacity to encounter the world from the perspective of others – results in people isolating themselves from reality, seeking shelter from the challenges and provocations of a plural, heterogeneous world, and thus becoming susceptible to authoritarian obedience to an unreality of slogans and ideologies.

As an empirical frame for the primarily theoretical argument in the chapter, the chapter explores experimental urbanism projects around the Tijuana–San Diego international border. I explore the political aesthetics of cross-border urban experiments that strive to open the border up to encounters with plurality and difference, and in so doing, to cultivate a richer, plural, unbordered "sense for reality". Through an extended discussion of an Arendtian account of cities, architecture, and aesthetics, I will suggest that experimental cross-border urbanisms can help to cultivate the sixth, "reality" sense by embracing a culture of exposure to plurality. In doing so, I make an argument for the significance of urban infrastructure as a form of *culture*, in the Arendtian sense, that endures, gathers, and expresses desire and fantasy.

First, I connect approaches to cities and aesthetics that are grounded in Arendt's ontology of appearance to recent debates within urban studies around methodological city-centricism. I suggest that an aesthetic approach can contribute to this debate by focusing on how social ontologies and boundaries are formed through the creation and contestation of aesthetic experience. I then explore insurgent urban practices and experimental urban design projects at the United States–Mexico border, exploring how these practices can be theorized as fabricating a temporary stage for marginalized groups, who suffer what Arendt calls the "insult of oblivion", to *appear* in the light of the public, and hence gain a fuller sense for reality. Next, I consider the role of materiality and infrastructure in the sense for reality, emphasizing that objects and infrastructures, and not only human actors, themselves "appear" on the political stage (Markell, 2011). This means objects and infrastructure, not just human performances, have an important role to play in establishing a sensory urban world that encourages exposure to plurality. Finally, exploring the role of judging in Arendt's political aesthetics, I highlight the importance of judgement, as an imaginative encounter with plurality, with contesting dominant urban border paradigms. I conclude that through judgement, experimental urban forms can be transfigured into places of *nonauthoritarian authority*.

For an Aesthetics of the City

"Aesthesis" refers to the presentation of the world to the senses, and aesthetic analysis foregrounds the importance of embodied sensory experience in the making of social relationships. However, a focus on aesthetics and the city contributes something more specific than a broad emphasis on bodies and sensations. One thing it adds is attention to the political relationship between the senses, sense, and the common. As Millner writes, "aesthetics can be thought of as the basis through which communities come to sense and feel 'in common'" (Millner, 2015: 76). Whereas an analytical emphasis on "the body" risks leading to an individualizing perspective, a focus on aesthetics requires asking how embodied sensation takes form and how it opens up or shuts down multiple perspectives, standpoints, and judgements. Aesthetic analysis refers to the constitution of the common through embodied judgements and evaluations. Aesthetic judgements are not merely individual or subjective feelings; rather, they are judgements that cultivate what Kant called an "enlarged mentality" by imaginatively occupying a multiplicity of perspectives. It is this *imaginative* aspect of judgement that makes aesthetics central to shaping the nature and possibilities of politics (Deutscher, 2016; Zerilli, 2005).

My suggestion is that an aesthetics of the city can add an important dimension to debates around the significance of "the city" in the epistemology of urban theory. As Blakey and Barron elaborate in more detail in the Introduction to this volume, the coherence of "the city" as an analytical category has been critiqued in several ways (Rickards et al., 2016). Brenner and Schmid (2014, 2015) question the outdated and unilluminating "methodological cityism" within urban studies, arguing that "the city" is a highly problematic unit of analysis in an age of planetary urbanization which is characterized by diverse forms of "concentrated, "extended", and "differentiated" urbanization. From this perspective, it is vital to move beyond any static, place-bound notion of the city, and instead to chart the diverse, global spatial processes that create neoliberal urbanization. This perspective of planetary urbanization foregrounds the "unevenly woven, restlessly mutating urban fabric of the contemporary world" (Brenner and Schmid, 2015: 170).

This provocation to urban studies is helpful in making the case for focusing attention on hinterlands, networks, and globalized processes. However, as a number of commentators have already observed, it risks minimizing situated knowledges and everyday lived experiences, perceptions, and politics, in which "the city" often plays an important part (Davidson and Iveson, 2015; McLean, 2018). It also imposes a totalizing perspective that discounts the contribution of feminist theory, queer theory, postcolonial theory, and diverse perspectives more generally, to urban studies, thereby imposing a singular, hierarchical epistemology of the urban. It falls back on a Marxist notion of capitalism as the "context of contexts" and the "epicentre" of urbanization processes, thereby discounting phenomena such as racialization as key organizing processes of many cities. In doing so, it relegates racism, patriarchy, heterosexism, and ablism to marginal positions within the constitution of the urban. Yet, the study of aesthetics and the city may offer an important corrective to this kind of hierarchical epistemology, shifting attention away from what the urban "is" or where its "epicentre" lies, towards the different ways in which cities are sensed and made sense of. Aesthetic analysis, by focusing on the constitutive role of sensation and judgement in urban experience, can uncover how certain urban processes and the city are foregrounded and others marginalized or silenced. In doing so, I will suggest that aesthetic analysis helps urban studies understand how new imaginaries of urban living as networked, globalised, and dispersed might be achieved.

An aesthetics of the city raises questions around beauty, taste, judgement, and value (Lehtinen, 2020), as well as the everyday sensuous life of buildings, landscapes, technologies, objects, commodities, and spaces. From a post-Kantian perspective, an aesthetics of the city raises the question of how *forms* of experience (i.e. the spatiality and temporality of experience) shape and are shaped by the urban environment, a question explored by Walter Benjamin amongst many others (Caygill, 1998). More than this, an aesthetics of the city asks how cities shape the formation of a "common sense" that distributes the allocation of roles, responsibilities, and the sensible itself (Rancière, 2004). The value of an aesthetics of the city in this context, I suggest, does not lie in its capacity to settle epistemological debates about methodological cityism or planetary urbanization. Rather, an aesthetics of the city can ask how and why the urban presents itself to experience in distinctive ways, shaping specific ways of inhabiting, sensing, and contesting a common horizon of action, thought, and judgement. An aesthetics of the city asks how urban forms, materials, and infrastructures gives shape and form to sensation, and what role the imagination has in sensing and making sense of the city. It focuses attention on the ways distinctive urban materialities shape structures of appearance and disappearance, and visibility and invisibility, such that some urban dwellers become visible but others are made invisible or inaudible. My suggestion in the remainder of this chapter is that an Arendtian theorization of the "sixth sense" in politics opens up a valuable set of questions for thinking about cities and urbanization processes.

The Sixth Sense: Exposure, Plurality, Reality

Amidst an ongoing re-evaluation of Arendt's writings on space, environment, nature, and architecture (Belcher and Schmidt, 2021: 4; Debarbieux, 2017; Dikeç, 2015; Last, 2017; Szerszynski, 2003; Teerds, 2022; Voice, 2013), I wish to explore the central but relatively undervalued importance of aesthetics for an Arendtian understanding of space, action, politics, and authority. Arendt's politics of aesthetics have been addressed by several writers. For example, Deutscher (2016) offers a useful reconstruction of Arendt's unfinished theory of (aesthetic) judgement. Honig (1993) provides a fascinating account of Arendt's debt to Nietzsche in her emphasis on beginnings, rebirth, the self as multiple, and Arendt's "virtuosic" theory of politics. This aesthetic account of politics rejects any effort to ground political authority in anything that is immune to contestation, amendment, or resistance, and instead looks to extraordinary actions that "disrupt the closures of moral and political systems and create new spaces of possibility" (Honig, 1993). Villa (1996), too, draws attention to Arendt's Nietzschean "aestheticism", suggesting that Arendt, like Nietzsche, embraces the aesthetic as a strategic response to the exhaustion of meaning in modernity. In a modern world not tied to transcendent foundations (such as sacred traditions, inherited authority, metaphysical truths, or any single principle determining how society should be ordered), the aesthetic offers "a way of rescuing the *possibility of meaning* in a nihilistic age" (Villa, 1996: 87). This means that the aesthetic is also the key to discovering a new kind of authority capable of creating the lasting, durable world that is a precondition of political action.

These accounts focus on the aesthetic in terms of actions that have no transcendent grounding, and demand evaluation according to aesthetic criteria such as "greatness" or "virtuosity" rather than epistemic criteria (truth) or ethical criteria (virtue, goodness). They have less to say, however, about the sensory constitution of politics, or to the role of art and culture in Arendt's thought. These issues are explored in studies by Curtis (1999) and Sjöholm (2015). According to Curtis, Arendt's theory explores ethical and political concerns *through*, not despite, her attention to the aesthetic. Aesthetic experience, she suggests, is not an analogy for political life, but "a central feature of it as a space of appearances" (Curtis, 1999: 15). Curtis argues that for Arendt, aesthetic, sensory experience is central to people's openness to reality – that is, their receptivity to and exposure to plurality. She shows that a "sense for reality" is a vital foundation of political life. Similarly, Sjöholm (2015) explores the role of the arts in cultivating a sensation of realness. Common to these works is a recognition that aesthetic experience and aesthetic judgement shape the forms of political action and authority in the modern context of unravelled traditions and foundations.

One issue that needs to be addressed in more detail, however, is the relationship between aesthetics, the sixth sense, and materiality. Arendt's thought has a notoriously difficult relationship with embodiment and materiality. Nevertheless, part of my aim is to take Arendt's though in new directions by connecting questions of aesthetics and the city to the materialities and infrastructures of urban spaces.

A key line of thought running through Arendt's political phenomenology is a critique of the loss of a feeling for reality in modernity. This decline in the experience of "realness" constitutes an impoverishment and devastation of experience – simultaneously a contributor to and symptom of the rise of totalitarianism. One factor behind this loss of a sense for reality is the building of societies designed, as Curtis (1999: 24) puts it, "to protect us from the plural heterogeneous quality of experience, from the intrusion of the surprising, the unusual, and the new". In such a world of borders, walls, segregation, defensive architecture, social media bubbles, echo chambers, and fetishized notions of "home" and nurturing domesticity, people are cut off from exposure to plurality, and are not called or challenged by others or events. We are anaesthetized to the sensory abundance of a varied, plural world. Indeed, sometimes anaesthetising feelings is an important way of refusing to inhabit depleting, damaging experiences (Bissell, 2022). One of Arendt's key insights into the politics of aesthetics and anaesthetics is that it is a sensitivity to and responsiveness to *plurality* – to the "encroachment of others" (Sjöholm, 2015) – that endows humans with a sense of reality. This sensitivity to plurality is aesthetic in nature.

A key idea in Arendt's work is that a sixth sense, which creates a sense of "realness", is an important condition for collective action. Arendt's guiding insight is that developing and maintaining a feeling of realness requires exposure to plurality. In *The Human Condition*, Arendt sets out a theatrical account of politics, highlighting the importance of a public realm or "space of appearance" where everything that appears in public can be seen, heard, and evaluated. Arendt contrasts three separate spheres of activity: labour (the part of human life devoted to subsistence, reproduction, and the fulfilment of biological needs); work (the fabrication of lasting and durable products, from tools and chairs to art); and action (non-purposive, self-contained acts which exhaust their full meaning in the performance itself). The paradigmatic space of action is the public realm, where political speech can be heard and responded to by a community of equals. The public realm is of paramount importance for creating a sixth sense, a feeling for reality, because appearance – being seen and heard by others and by ourselves – "*constitutes*" reality (Arendt, 1998: 50). Thus, whilst political attention to inequality and exclusion tends to focus on "biological" privation – lack of access to food, shelter, services, education, and security – Arendt draws attention to an even more fundamental privation, which is a lack of the capacity to appear, and hence, expulsion to a domain of unreality, where neither the self nor others feel real. Even the intensities of "private" experiences – for example, sensory experience, passion, pain, and thought – only acquire a full reality when they are deindividualized and shaped into something suitable for public appearance. Thus, "the presence of others who see what we see and hear what we hear assures us of the reality of the world and ourselves" (Arendt, 1998: 50). This sense of reality contrasts with the alienation and repetitive rhythms of labour, which Arendt argues produce "a withering sense of irreality" (Sjöholm, 2015: 107).

From an Arendtian perspective, an urban environment conducive to politics is one that protects its inhabitants' exposure to risk, plurality, and uncertainty. For this reason, it requires risky political spaces where courage is needed to expose oneself in the light of the public. Yet the way cities are designed often reflects a "great, unreckoned fear of exposure" (Sennett, 1990: xi). The growing segregation of cities into security enclaves of gated communities, ghettos, and slums, reflects an emphasis on excluding what privileged classes and ethnic groups experience as discomforting otherness. Gated communities, for example, meet the desire for security and safety by excluding signs of "otherness" (Bickford, 2000; Teerds, 2022). Urban segregation, enforced through various forms of "interdictory" urban design, gentrification, exclusion, and stigma, is detrimental to the exercise of the "sixth", community sense. To some extent, this affects both those "inside" and "outside" these divisions, but it is intensified for those on the "inside". Those on the outside, the excluded, have greater exposure to reality than the privileged class on the inside, since they are not cocooned within sanitized interior or quasi-interior spaces (or what Sloterdijk (2013) calls the global hothouse or the world interior of capital). This is why Arendt saw great political potential in figures such as the pariah and the exile (Dubow, 2020).

Communities living on different sides of an exclusionary wall, deprived of a shared space of appearance, cannot be expected to share the same reality. People who do not share the same reality cannot truly "appear" to each other or be exposed to each other's performances, words, and actions. The result is urban landscapes that insulate the privileged from the encroachment of others who are unlike themselves. They insulate them from a diversity of perspectives, opinions, and sensations. In doing so, these urban landscapes shelter people from a discomforting exposure to others, and deprive those same people – as well as those whom they exclude – of a sense for reality.

What sorts of urban landscape might encourage, rather than prevent, the emergence of plurality of perspectives and a pluralising space of appearance? One place where Arendt directly addresses spatial phenomena is her observation that monuments can help create a durable world: they are marked off from the common world of everyday practices, and provide people with a shared point of orientation in an intensified place of meaning. In addition, monuments are modes of action, a kind of political speech. The problem is that they are a form of political speech that easily enforces a singular, stable meaning (thereby denying, rather than maximising, plurality). Whilst it is true that all monuments can be and are read in different ways by different groups of people, many monuments still typically aspire to create singular, universalizing, and easily legible meanings. By contrast, a monument that truly *politicizes* the landscape, in an Arendtian sense, would be a monument that recognizes and amplifies plurality by staging and foregrounding multiple perspectives, and opening up a public sphere for those multiple perspectives to be voiced and heard. Similarly, a monument can become politicized when it becomes a focal point of multiple, diverse readings, as can be seen with the recent toppling of various colonial statues such as the Colston statue in Bristol.

Arendt's thought can also be connected to the politics of landscape. Following her division of human activity into labour, work, and action, we can think of landscape in each of these guises. First, landscapes may be shaped by labour, as famously discussed by Ingold in his account of Bruegel's *The Harvesters* and the temporality of landscape (Ingold, 1993). Debarbieux (2011) observes how landscape representations in 18th and 19th century Europe often focused on peasants' labour, aiming for a realist representation of the landscape, but typically idealizing notions of a supposed harmony between labourers and nature. Second, landscapes may be produced as work, where labour is subordinated to the production of representation. Debarbieux gives the example of the landscape designs of Capability Brown, though we might equally consider the various monumental urban landscapes that confer a sense of durability upon dominant social forms and ideals. Finally, landscapes may be produced through action, where landscapes are created as a form of collective and participatory practice. Such landscape projects

> link politicians, professionals, and residents in a reflection on what the landscape represents in what all of them think they have in common. They also testify to the capacity of landscape today to represent a point of commonality, at a time when cultural identities ... show—indeed, exalt—diversity and heterogeneity.
>
> (Debarbieux, 2011: 144)

This is an important point, for it raises an important question: how might urban landscapes have a *politicising* function, in the sense of introducing disruptive new voices, perspectives, and appearances that transform the stage upon which appearances are staged and demands are voiced? In asking this question, I raise the question of what the political is and how it can be defined. Against Arendt's frustratingly formalist sense of the political, which can seem to carefully police a boundary between the political as a space of exchange of opinion and the "private" realm of materiality, corporeality, and need, my discussion of participatory kinds of urban landscape production in the following section will highlight how material urban practices can themselves redefine and restage the nature, terrain, and borders of the political sphere.

Improvising an Infrastructure of Appearance

On the outskirts of Tijuana, in the dead of night, a group of hundreds of urban activists make their way to a plot of unused land (usually, but not always, publicly owned) on the hills in the east, and start constructing informal housing with carefully chosen materials: used tyres, concrete blocks, corrugated metal, used garage doors, and packing crates. Much of this material is discarded material from over the border in San Diego, brought down wholesale by local contractors. Occasionally entire prefabricated houses, scheduled for demolition in San Diego, are bought at low prices and integrated into the informal settlement. The urbanism here thus dramatizes the inequalities of national bordering: valuable commodities flow north of the border, whilst discarded items flow south (Cruz, 2005).

Due to the small amount of space available, these prefabricated houses are often placed on top of simple single-story frames, leaving an empty ground floor level that can be used for shops, car repair, or social space. Building the housing is a race against time: once a settlement exists, Mexican law does not allow it to be demolished, and the government must eventually start supplying it with electricity, water, and roads. This means the settlements must be "completed" overnight, as if by magic. However, they are then continuously improved, adapted, and expanded over a period of several months. The result of this process is a kind of informal housing that is inventive, adaptable, and resourceful. The informal urbanism here reflects the contradictory nature of the United States–Mexico border at a site where 60 million people cross every year, moving vast quantities of goods across the border. The border, described by Gloria Anzaldúa (1987: 3) as an open wound in which first and third world meet, is continually defined and redefined in the unstable balance of hugely unequal, powerful forces: on the one hand, the militarized government urban policy that transforms San Diego into a vast gated community; and on the other, the informal, insurgent, and "illegal" actions that proliferate in both border cities in spontaneous occupations and appropriations that transgress the wall dividing the two communities (Cruz, 2005: 33).

An architectural project led by Teddy Cruz and Fonna Forman, called Manufactured Sites, deploys a technique of "urban acupuncture" to work with, facilitate, and augment these insurgent urbanisms. It involved the design of a simple aluminium structure to act as a frame for recycled materials. A key political part of the project is to demand that maquiladoras (vast, exploitative factories on the Mexican side of the border taking advantage of cheap Mexican labour and proximity to US markets) make and distribute these frames to the local population (Cruz, 2005). These aluminium structures are cheap and simple to manufacture and can be used as frames on which to build the recycled materials, thereby making the dwelling safer, longer lasting, and more secure. The frames are distributed during the overnight incursion, and families can use them to quickly make their first traces of occupation on the land. The frames have prefabricated footings that can be injected into the walls of rubber tyres, adding stability. The frame comes with a clip-on refillable water tank which can hold two weeks' supply of water. The frame is used as a hinge that facilitates the connections between various recycled materials. A manual is provided that helps people attach commonly used recycled materials such as pallets, recycled joists, and rubber tyres.

The Manufactures Site project creates aesthetic links between activism (campaigning against exploitative labour practices), urbanism (designing infrastructure to facilitate rapid, stable, and durable safe informal urban infrastructures), and art (making a project whose visually striking images and landscapes have been displayed in elite global galleries, thereby creating new kinds of public for these experimental infrastructures). In doing so, it augments the practices of insurgent urbanism by witnessing and amplifying the actions of marginalized political actors. It amplifies the existing practices of marginalized communities, strengthening the infrastructure through which marginalized groups assert the authority and dignity to speak as "improper" political actors. It attempts to reconfigure a space of hidden and illegal activities as a reconstituted political stage (Drexler, 2007). Reading with Arendt, I suggest, we can describe this kind of improvised infrastructure as co-constructing a "stage of appearance" whose borders are not fixed, like the Ancient Greek polis, but which challenges the borders and boundaries between the political and the nonpolitical.

In Arendt's writing, the city plays an important thematic role, even though she rarely discusses specific architectures, landscapes, or urban forms. Political space is conditioned, even made possible, by walls and buildings that are created for a community. In her own brief comments about architecture in *The Human Condition*, Arendt draws parallels between architecture and law, suggesting that they both define and protect the political realm. The key point here is that they are not themselves political, because it is the people themselves who constitute the political sphere: "these tangible entities themselves were not the content of politics (not Athens, but the Athenians, were the polis)" (Arendt, 1998: 194). From this perspective, the Manufactured Landscapes project does not define a politics, but co-constructs a *stage* upon which political actions can be performed.

Yet is the creation of the space – the temporary and provisional coming together of actors to assemble an urban infrastructure that makes demands and provocations – not itself a political act? This is a key part of my argument. Teerds (2022) observes a tension in Arendt's thought. He notes that whilst Arendt seems to limit action to formal public spaces, this may contradict her argument that appearances (and thus action and speech) precede the establishment of the public realm: as Arendt puts it, "[t]he space of appearance comes into being wherever men are together in the manner of speech and action, and therefore predates and precedes all formal constitution of the public realm and the various forms of government, that is, the various forms in which the public realm can be organised" (Arendt, 1998: 199). If we fully take this argument on board, the political dimension is opened to all spaces where people can encounter one another: not just spaces assigned to political speech, but to everyday spaces. All that is required is the presence of "others", for there is no appearance without encountering others. Any kind of everyday space harbours the possibility "that one day, an everyday encounter leads to public appearance, and that it becomes a space, not only allowing multiplicity, but asserting plurality" (Teerds, 2022: 9). Teerds is getting at a key difference here between a mere coincidence of many different actors and opinions (*multiplicity*) and the existence of a shared common space in which actors appear *to each other* and their appearances are witnessed and judged *(plurality)*. Another way of putting this is that Arendt's preoccupation with establishing the boundaries of the political is in tension with an understanding of action as a miraculous, unpredictable new beginning, which can take place anywhere. The unpredictability of action makes it impossible to bound it in advance, yet Arendt seems preoccupied with doing so (Honig, 1993).

In fact, in modernity it is plausible that the formation of a rich "sixth sense", a sense for reality, may only really be possible from positions of marginality and exclusion, such as the "pariah" (Dubow, 2020). Such practices of freedom do not arise from equal access to a shared public sphere, but by improvising new, temporary spaces of political speech and acts through which claims can be voiced, heard, evaluated, and judged. Such spaces of appearance need not constitute a permanent, carefully demarcated sphere of human sociality, but may be built as something more ephemeral and transitory, outside the formal public sphere, by disenfranchised groups. "Excluded from 'normal' politics, these groups find ways of constituting spaces of appearance in which their own worldly concerns and practices of freedom can make themselves felt" (Curtis, 1999). Existing outside formal public realms, such political subjects must create *their own* stages upon which to appear. Such practices literally build their own stage and create their own audience at the *limit* of the visible. In the absence of secure political foundations, the stage must be reinvented. Marginalized communities must continually recreate their stage, each time making their entrance anew (cf Nancy, 2000).

Landscape, World, and Materiality

What is not yet clear in this account of the urban "sixth sense" is the role of materiality and sensory experience in the formation of a sense for reality. Arendt is notoriously suspicious of the encroachment of matter on human political affairs (Last, 2017). Yet, her accounts of world and worldlessness also offer a passionate defence of the importance of "things" for creating the element of durability within human affairs without which the performative nature of action could have no meaning, no stage from which to evaluate it, and to prevent it from immediately disappearing into nature's ceaseless flux. Things and materials, within her thought, are vitally important for allowing action to acquire meaning and weight. Without a stage, a durable world, upon which to appear, a rich sense for reality is impossible.

One of the worst aspects of poverty, from an Arendtian perspective, is world alienation or "loss of world": a loss of a durable material structure to provide safety, privacy, and a relationship with others (Luttrell, 2015). The circumstances of the 24% of the world's population who live in slums (UN, 2021) offer one clear example of this kind of alienation. Indeed, this loss is perhaps the most distinctive characteristic of modernity: "World alienation, and not self-alienation as Marx thought, has been the hallmark of the modern age" (Arendt, 1998: 254): not a loss of self, but a loss of objectivity, of material connectivity. A "world" is created through durable, material infrastructures and products: buildings, institutions, roads, cities, monuments, laws, and artworks. It is also closely connected to a distinctive form of social coordination: authority (Honig, 1993). When humans belong to a world, they are sheltered from the violence of nature and are related to each other. World is a material form that creates an experience of time that is central to a sense for reality: an experience of duration, of temporal depth. Without durable material structures to relate people together in time, there is no space for freedom as a miraculous beginning that sets an unpredictable chain of events in motion. Action, the creation of a miraculous new beginning, requires a stable temporality as its background: without this, it would be indistinguishable from labour, with which it shares a fleeting, processual temporality. Living together in the world means that human relations are mediated by things which create an "in-between" that at once separates and relates people (Arendt, 1998: 52). This stability is one of the conditions for freedom to appear. When people are in a world, they can reveal themselves as unique beings, in a temporality which is neither one of repetition nor progress, but of new and unexpected beginnings (Luttrell, 2015: 875). Worldlessness or world alienation, on the other hand, corresponds to an anaesthetics of invisibility: it consists of being excluded from the "public" realm of visibility and confined to the realm of biological reproduction. Those who suffer the condition of worldlessness are condemned to the "insult of oblivion" (Curtis, 1999). They lack reality itself, because when they lack a world, they are deprived of the space "in-between" that at once distinguishes them as an individual and relates them to others.

Yet, matter, materials, and infrastructures are not the mere backdrop to human appearances. In *The Life of the Mind*, Arendt's account of individuation makes it clear that appearance is not merely a human issue, but applies to everything that exists. Arendt's account of individuation collapses the ancient metaphysical hierarchy of Being over Appearance. Against this metaphysical tradition, Arendt affirms that "to be" simply is "to appear". Because the idea of "appearing" presupposes the existence of others to appear *to*, this means that being is intrinsically plural. Plurality is the condition of singular being.

> Being and Appearing coincide. Dead matter, natural and artificial, changing and unchanging, depends in its being, that is, in its appearingness, on the presence of living creatures. Nothing and nobody exists in this world whose very being does not presuppose a spectator.
>
> (Arendt, 1978: 19)

In this way, Arendt's account of appearance implies a kind of aesthetic vitalism. "Could it not be", she asks, "that appearances are not there for the sake of the life process but, on the contrary, that the life process is there for the sake of appearances?" (Arendt, 1978: 27).

It is apparent, then, that not only humans, but also nonhuman life and inanimate matter all *appear*, and their reality derives from their capacity to appear to others. This might sound radically anthropocentric or "correlationist" – and Arendt's humanism does indeed presuppose fundamental differences between human and nonhuman appearances – but it should be observed that Arendt is clear that appearing doesn't presuppose human spectators. A bird's reality, for example, might be guaranteed by its appearence to other birds. Still, this ontology of appearance raises some problematic questions. In a world of appearances – in the cycle of appearing, spectating, and judging that inevitably brings with it the risk of error and deception – how can reality be guaranteed? Arendt (1978) suggests this feeling of realness arises from a threefold commonness. First, the five senses can be united by having the same object in common. Second, members of a species may share a "context" that endows every object with its specific meaning. Third, all other living creatures endowed with sense may observe the object from entirely different perspectives, but can all agree on its basic identity: they all encounter the same object. Thus, for Arendt, "Out of this threefold commonness arises the *sensation* of reality" (Arendt, 1978: 50).

In this passage, the notion of "context" does a lot of work, and is not clearly explained by Arendt. But what she seems to have in mind is, once again, the worldly materials, objects, infrastructures, and things that both separate and relate everything that "appears" to others. When people do not share the same context, they no longer share the same reality and are, thus, cut off from exposure to plurality. Fractured human societies, cut off from each other by walls and borders, risk being deprived of a shared context, of shared exposure to others, and thus of a rich sense of reality. Material infrastructure provides those common objects that are required to share a world. But here another distinction needs to be made between the kind of things that have a wholly functional, practical use, and those that have a use that transcends function and which thereby achieve greater durability. This is precisely what Arendt defines as *culture*: "[a]n object is cultural to the extent that it can endure; its durability is the very opposite of functionality, which is the quality which makes it disappear again from the phenomenal world by being used and used up" (Arendt, 1961: 208). Arendt approvingly recalls the Ancient Greek idea of culture as "the mode of intercourse of man with the things of the world" (Arendt, 1961: 213). Thus, culture is material, but it also goes beyond its materiality in its capacity to endure, unlike functional objects.

These reflections on the importance of appearances in the world of things, I suggest, shed new light on the importance of experimental urban infrastructures in the constitution of an Arendtian political aesthetics. The Manufactured Sites project is one of many architectural attempts to harness the productive capacities of self-help housing (Ward, 2019). One thing that is striking in this project is its focus on infrastructure as part of a dynamic assemblage of materials, buildings, borders, practices, and goods. Approaching this practice from the perspective of political aesthetics demands attention not only to the purely technical functioning of this infrastructure, but also to how it functions as a form of culture. Such a "poetics of infrastructure" requires asking, for example, how infrastructural materials "emerge out of and store within them forms of desire and fantasy and can take on fetish-like aspects that sometimes can be wholly autonomous from their technical function" (Larkin, 2013: 329).

Amin (2014), writing of a similar practice of informal housing occupations in northeast Brazil, exemplifies an aesthetic approach to the world-building aspects of insurgent urbanism. He observes how,

> These hyper-visible and constantly evolving infrastructural developments make the atmosphere of place that forms the precognitive of mental, sensory and affective dispositions: the residents' experience of living in the settlement, their feelings and obligations towards each other, their attachment and responsibility towards shared public spaces, their expectations from the commons.
>
> (Amin, 2014: 146)

For Amin, collectively improvised infrastructural practices do not merely provide the background or "context" to public appearances, but actively shape the feelings, affects, and relations that help to structure the possibilities of public space. Such infrastructural practices do not only create functional use-objects, but also help to *narrate* and *affectively condition* a community – they help a community to make sense of itself, and to present itself to others.

The same observation can be made about the urban incursions we see in Tijuana. Infrastructural materials here are not merely a stable backdrop to political appearances. The infrastructures themselves *appear*. Arendt is clear that all things, insofar as they take shape, appear. For most things, appearance is subordinate to function: to what they are intended to be used for. Artworks are the only special kind of thing which are produced solely for the purpose of appearance. Yet it is precisely this distinction between the functional and the non-functional that is troubled by the Manufactured Sites project, where the everyday is transfigured into artwork, and where art is revealed as a practice of collective self-realization. The Manufactured Sites project is precisely a *cultural* project because it narrates a specific mode of intercourse between people and the things of the world. The design and aesthetics of the housing, where residents design housing and infrastructure that is high-density, flexible, multi-use, and adaptive, practices a kind of narrative world-making in two key ways. First, the injection of new urban infrastructural materials such as the aluminium frame enables the construction of urban settlements that are more durable than they could otherwise be. In this way, the project contributes to the world-building practices of informal urban communities by helping improvised and fragile world-making materials become more durable, and hence to shift from labour (meeting cyclical biological needs) to work (constituting a durable *world*).

Second, and perhaps more importantly, the Manufactured Sites project contributes to world-building by taking shape as an artwork that makes visible and narrates the flow of objects across the border. The recycled objects brought to Mexico from the other side of the border acquire a density of meaning, speaking to the inequities of international bordering practices, and demonstrating the extent to which communities on either side of the border are still shaped by each other, despite the attempts to separate them through violent militarized bordering practices. Architecture is never neutral, and through its forms architecture can revive memories, evoke narratives, and provoke expectations (Teerds, 2022). In this case, a world is evoked which is not restricted to the informal housing community itself but evokes a commonality that relates communities on either side of the border, divided as they are by physical, economic, legal, cultural, material, and social infrastructures. By emphasizing the global nature of the recycled materials and the economic forces that produce them and the need for them, as well as the broader social problems facing marginalized communities on either side of the border, the architectural practices of Estudio Teddy Cruz make an explicitly politicized attempt to narrate a cross-border world. In doing so, they enact a politics of aesthetics, nurturing an imaginative "sixth sense", a sense of *reality*, a sense of a common, shared world available to shared opinion and judgement.

Judging

There is one more way in which the Manufactured Sites project, and the urban incursions of which it is a part, intervenes in an aesthetics of the city: by shaping the sense for reality. Through its practices of border-transgressing world-making, the project invites new practices of *judgement* that, for Arendt, constitute perhaps the most important contribution of aesthetics to politics. In the Kantian tradition, judgement is the faculty that enables people to order and make sense of experience. Kant saw judgement as the ability to think the particular as contained under the universal – for determining whether something stands under a given rule or not. It is through judgement that objects come into being through recognition of a concept. Arendt believes Kant's greatest discovery in the *Critique of Pure Reason* to be the importance of imagination for the cognitive faculty (Arendt, 1982: 80). However, it is the *Critique of Judgment* that contains Kant's true political philosophy. Kant's key insight here is that taste involves a form of "community sense" or representative thinking. Taste is not subjective, but not determined by community either; it is connected to community but not overwhelmed by it. Taste is neither subjective nor objective. Instead, taste is an individual faculty that presupposes the presence of others, and requires an imaginative capacity to encounter reality from the diverse standpoints of heterogeneous others.

For Arendt, the imperative of the sense for reality, or what she also calls the "community sense", is to think from the standpoint of others. This is a practice through which we "enlarge" our mentality, and shed off the individual, private, and incommunicable aspects of experience. Whereas Kant presupposes some kind of *a priori* community sense, Arendt appeals to the ongoing activities of community: "when one judges, one judges as a member of a community" (Arendt, 1982: 72). This is a question of presenting others by travelling in imagination. We travel to "the place where they stand, the conditions they are subject to, which always differs from one individual to the next, from one class or group as compared to another" (Arendt, 1982: 43). This is not a question of empathy – of feeling another. Rather, as Curtis argues, we present the encroachment of others in ways that augment the self.

> In representative thinking, the other's view becomes a dimension, a world. It encroaches on mine, I am extended, I see what the world looks like from this particular spot ... Representative thinking responds to and brings forth the essential dimensionality of the world.
>
> (Curtis, 1999: 118)

It is the imaginative capacity to occupy the perspectives of others that ultimately confers dimensionality and a feeling of realness in the world.

Judging is a faculty that creates pleasure in the contingencies of particular, plural appearances. This pleasure relates directly to the experience of freedom, because the only way to affirm human freedom as a miraculous "beginning" in the world is to take pleasure from singular, unique practices of freedom, rather than feeling freedom as an intolerable weight (Beiner, 1997). Judgement is the faculty that makes possible this joy in the particular and the singular. In doing so, it acquires the weight of the durable. The importance of judging in Arendt's thought lies in its capacity to augment the capacity to begin. Judgements, when emerging from an imaginative encounter with plurality, impart a kind of durability to the world, because they are anchored in a sense of reality that encompasses multiple perspectives, all linked and separated by a common stage, a shared reality. In this way, judgements confer a certain weight and durability on the world. Of course, not all judgements achieve this – judgement is too often anchored in prejudice and singularity. However, a form of aesthetic judgement based on imaginative encounters with multiple perspectives and opinions may have this capacity to anchor the world and make it durable.

Might it be, then, that the most important task for a democratic urbanism is to cultivate the capacity for judgement – that is, to boost the powers of the imagination by amplifying our sensitivity to the encroachment of others? This is a particularly important issue for those wealthy segments of society who shelter behind walls and within bubbles, and are thus especially liable to see their imaginative "sixth sense" decay. The urban projects of Estudio Teddy Cruz all make visible and problematize the nature and function of the urban border. Dramatizing the unequal flow of commodities, objects, bodies, dreams, and nonhuman life, these projects attempt to stage an encounter with multiple perspectives and viewpoints. Whether injecting durable infrastructural materials into fragile improvised informal housing, or else contesting US zoning laws that prioritize low-density urban sprawl and forbid housing that meets the cultural practices of Latinx communities, Cruz's projects produce a form of border-crossing *culture*, expressing the possibility of new modes of intercourse between people and the things of the world. They design modes of intercourse that link communities in new ways, imagining new forms of citizenship and affirming the possibilities of improvisatory, flexible, adaptable architectures. These architectures take advantage of existing flows of materials, leverage pressure to transform those flows, and embrace forms of porosity that encourage exposure to multiple viewpoints, perspectives, and attitudes – and so, to the cultivation of a richer sixth sense for reality.

Conclusion: Towards a New Urban Landscape of Nonauthoritarian Authority?

Sennett (1990) suggests that many modern urban landscapes are produced in ways that have lost sight of "the outside" as a dimension of human experience. Instead, they fall back into the comfortable security of interiority, eliminating the possibility of exposure to difference and plurality. In modernity, he suggests, the landscapes that generate an individualist sense of separation and isolation can stimulate a pleasurable (but alienated and impoverished) experience of authority. He discusses how the Rockefeller Center complex of skyscrapers in New York creates a sense of protection and security through its very emptiness and precision (Sennett, 1990). Such highly rationalized landscapes promise people a safe distance from others but also connect them to the march of progress and the pull of a rationally perfectible future.

For Arendt, the loss of authority in modernity is closely connected to the decline of tradition and sacred foundations. These threaten to leave experience anchorless or empty. Authority was a form of social and political coordination that helped create depth in human experience, serving to confer longevity, and durability on political worlds. It helped to ground a sense of reality by linking experience to a thread that led back to the great, beautiful deeds of immortals. Authority was conferred to people or institutions tasked with protecting and amplifying these sacred foundations. However, Arendt had no intention of resurrecting this kind of authority – she is hardly an apologist for Ancient Rome. She explicitly affirms her participation in the destruction of metaphysics, and thus the destruction of the Roman model of authority that was passed down to European societies for centuries. Such authority cannot protect freedom, since it grounds experience in a single foundation rather than exposure to plurality. Yet, Arendt suggests the loss of authority does not mean that people have lost their ability to care for the world. The key task is to think through the possibilities for caring for a world when one important model for doing so (foundational, metaphysical authority) is broken.

This search for a new form of modern non-authoritarian authority that is dynamic, revisable, temporary, and mutable is a crucial issue for contemporary politics. My suggestion is that experimental urbanism can play an important part in this search. Improvisatory experiments with urban infrastructure hold the potential to boost the capacity for judgement that underpins the "sixth sense" that connects people to others (both human and nonhuman) and to the world. Experimental urbanism can amplify the powers of the imagination and judgement. In doing so, they can help invent new kinds of nonauthoritarian authority. This is because judgements command a kind of authority: they command an obedience in which we retain our freedom (Arendt, 1961). The gravity of metaphysical authority can be reinvented as the weight of forms of nonauthoritarian judgement which affirms the human capacity to evaluate particulars not by merely responding with "subjective" reactions but by using imagination to inhabit multiple perspectives and to be transformed by them. Judgement is not the triumph of subjectivism, but the opposite. Judgement presupposes a kind of submission, where the existence of other viewpoints inhabits the subject's perspective to the extent that she is encroached upon, forced to try to take up new standpoints, and find new points of view (Sjöholm, 2015).

In this light, Cruz's (2005) border-crossing urban designs can be seen as a new landscape of nonauthoritarian authority. In contrast to dominating landscapes of authority which naturalize and generalize the single, partial viewpoints of the powerful, Cruz's experimental urban landscapes might instead be thought of as urban landscapes of *non-authoritarian* authority: landscapes which create new kinds of durability, beauty, and worlding through practices that stage encounters with plurality. Unlike militarized border practices that shut off difference and insulate people from exposure to plurality, these urbanisms dramatize the possibility of creating urban forms that encourage imagination, judgement, and a richer "sixth sense" of the city. This sixth sense, the reality sense, is what helps people avoid alienation from world and earth (Belcher and Schmidt, 2021). It is the sense of participating in a shared world, a common reality. It is a *sense* akin to the Aristotelian *sensus communis* in that it is not reducible to immediate sensation or rational thought, but cooperates with both. Unlike the Aristotelian and Kantian notions of *sensus communis*, however, the sixth sense is a sense of existing within a world of shared appearances, actions, and performances of freedom. Experimental border-crossing urbanisms such as the ones discussed here can help to make a palpable, shared, common world of exposure to others and imaginative judgements of plural, sometimes disruptive stages of appearance.

References

Amin, A. (2014). Lively infrastructure. *Theory, Culture & Society*, *31*(7–8), 137–161. https://doi.org/10.1177/0263276414548490

Amin, A., & Richaud, L. (2020). Stress and the ecology of urban experience: Migrant mental lives in central Shanghai. *Transactions of the Institute of British Geographers*, *45*(4), 862–876. https://doi.org/10.1111/tran.12386

Anzaldúa, G. (1987). *Borderlands: La frontera*. San Francisco, CA: Aunt Lute.

Arendt, H. (1961). *Between past and future: Six exercises in political thought*. New York: Viking.

Arendt, H. (1978). *The life of the mind*. San Diego, CA: Harcourt.

Arendt, H. (1982). *Lectures on Kant's political philosophy*. Chicago, IL: University of Chicago Press.

Arendt, H. (1998). *The human condition*. Chicago, IL: University of Chicago Press.

Arendt, H., & Kroh, J. (1964). *Eichmann in Jerusalem*. New York: Viking.

Beiner, R. (1997). Rereading Hannah Arendt's Kant lectures. *Philosophy & Social Criticism*, *23*(1), 21–32.

Belcher, O., & Schmidt, J. J. (2021). Being earthbound: Arendt, process and alienation in the Anthropocene. *Environment and Planning D: Society and Space*, *39*(1), 103–120. https://doi.org/10.1177/0263775820953855

Bickford, S. (2000). Constructing inequality: City spaces and the architecture of citizenship. *Political Theory*, *28*(3), 355–376. https://doi.org/10.1177/0090591700028003003

Bissell, D. (2022). The anaesthetic politics of being unaffected: Embodying insecure digital platform labour. *Antipode*, *54*(1), 86–105. https://doi.org/10.1111/anti.12786

Brenner, N., & Schmid, C. (2014). The 'urban age' in question. *International Journal of Urban and Regional Research*, *38*(3), 731–755. https://doi.org/10.1111/1468-2427.12115

Brenner, N., & Schmid, C. (2015). Towards a new epistemology of the urban? *City*, *19*(2–3), 151–182. https://doi.org/10.1080/13604813.2015.1014712

Buck-Morss, S. (1992). Aesthetics and anaesthetics: Walter Benjamin's artwork essay reconsidered. *October*, *62*, 3–41.

Burks, V. (2002). The political faculty: Arendt's 'Ariadne Thread' of common sense. *Theory & Event*, *6*(1). https://dx.doi.org/10.1353/tae.2002.0002

Caygill, H. (1998). *Walter Benjamin: The colour of experience*. London: Routledge.
Cowan, A., & Steward, J. (2007). *The city and the senses: Urban culture Since 1500*. Farnham: Ashgate.
Cruz, T. (2005). Tijuana case study. Tactics of invasion: Manufactured Sites. *Architectural Design*, *75*(5), 32–37. https://doi.org/10.1002/ad.133
Curtis, K. (1999). *Our sense of the real: Aesthetic experience and Arendtian politics*. Ithaca, NY: Cornell University Press.
Davidson, M., & Iveson, K. (2015). Beyond city limits: A conceptual and political defense of 'the city' as an anchoring concept for critical urban theory. *City*, *19*(5), 646–664. https://doi.org/10.1080/13604813.2015.1078603
Debarbieux, B. (2011). The political meaning of landscape (through the lens of Hannah Arendt's *The Human Condition*). In *The place of landscape: Concepts, contexts, studies*. (C. Salzani, Trans.), edited by J. Malpas. (pp. 131–149). Cambridge, MA: MIT Press.
Debarbieux, B. (2017). Hannah Arendt's spatial thinking: An introduction. *Territory, Politics, Governance*, *5*(4), 351–367. https://doi.org/10.1080/21622671.2016.1234407
Deutscher, M. (2016). *Judgment after Arendt*. London: Routledge.
Diaconu, M., Heuberger, E., Mateus-Berr, R., & Vosicky, L. M. (2011). *Senses and the city: An interdisciplinary approach to urban sensescapes*. Münster: LIT Verlag.
Dikeç, M. (2015). *Space, politics and aesthetics*. Edinburgh: Edinburgh University Press.
Drexler, J. M. (2007). Politics improper: Iris Marion Young, Hannah Arendt, and the power of performativity. *Hypatia*, *22*(4), 1–15.
Dubow, J. (2020). *In exile: Geography, philosophy and Judaic thought*. London: Bloomsbury.
Gassner, G. (2019). *Ruined skylines: Aesthetics, politics and London's towering cityscape*. London: Routledge.
Goonewardena, K. (2005). The urban sensorium: Space, ideology and the aestheticization of politics. *Antipode*, *37*(1), 46–71.
Hayles, N. K. (2007). Hyper and deep attention: The generational divide in cognitive modes. *Profession*, *2007*(1), 187–199.
Honig, B. (1993). *Political theory and the displacement of politics*. Ithaca, NY: Cornell University Press.
Ingold, T. (1993). The temporality of the landscape. *World Archaeology*, *25*(2), 152–174.
Kingsbury, P. (2010). Unearthing Nietzsche's bomb: Nuance, explosiveness, aesthetics. *ACME: An International E-Journal for Critical Geographies*, *9*(1), 47–61.
Larkin, B. (2013). The politics and poetics of infrastructure. *Annual Review of Anthropology*, *42*(1), 327–343. https://doi.org/10.1146/annurev-anthro-092412-155522
Last, A. (2017). Re-reading worldliness: Hannah Arendt and the question of matter. *Environment and Planning D: Society and Space*, *35*(1), 72–87. https://doi.org/10.1177/0263775816662471
Law, L. (2001). Home cooking: Filipino women and geographies of the senses in Hong Kong. *Ecumene*, *8*(3), 264–283.
Lehtinen, S. (2020). Editorial introduction to the special volume on urban aesthetics. *Contemporary Aesthetics* (*8*). https://contempaesthetics.org/2020/07/16/editorial-introduction-to-the-special-volume-on-urban-aesthetics/
Luttrell, J. C. (2015). Alienation and global poverty: Arendt on the loss of the world. *Philosophy & Social Criticism*, *41*(9), 869–884. https://doi.org/10.1177/0191453714567734
Markell, P. (2011). Arendt's work: On the architecture of *The human condition*. *College Literature*, *38*(1), 15–44.
McLean, H. (2018). In praise of chaotic research pathways: A feminist response to planetary urbanization. *Environment and Planning D: Society and Space*, *36*(3), 547–555. https://doi.org/10.1177/0263775817713751

Millner, N. (2015). Activist pedagogies through Rancière's aesthetic lens. In *Geographical aesthetics: imagining space, staging encounters*, edited by H. Hawkins & E. Straughan (pp. 71–90). Farnham: Ashgate.

Mubi Brighenti, A., & Pavoni, A. (2019). City of unpleasant feelings. Stress, comfort and animosity in urban life. *Social & Cultural Geography*, *20*(2), 137–156. https://doi.org/10.1080/14649365.2017.1355065

Nancy, J.-L. (2000). *Being singular plural*. Stanford, CA: Stanford University Press.

Paterson, M. (2007). *The senses of touch: Haptics, affects and technologies*. Oxford: Berg.

Rancière, J. (2004). *The politics of aesthetics: The distribution of the sensible*. (G Rockhill, Trans.). London: Continuum.

Rickards, L., Gleeson, B., Boyle, M., & O'Callaghan, C. (2016). Urban studies after the age of the city. *Urban Studies*, *53*(8), 1523–1541. https://doi.org/10.1177/0042098016640

Sennett, R. (1990). *The conscience of the eye: The design and social life of cities*. New York, NY: Knopf.

Sjöholm, C. (2015). *Doing aesthetics with Arendt: How to see things*. New York, NY: Columbia University Press.

Sloterdijk, P. (2013). *In the world interior of capital: For a philosophical theory of globalization* (W. Hoban, Trans.). Cambridge: Polity.

Stiegler, B. (2010). *Taking care of youth and the generations*. Stanford, CA: Stanford University Press.

Szerszynski, B. (2003). Technology, performance and life Itself: Hannah Arendt and the fate of nature. *The Sociological Review*, *51*(2), 203–218. https://doi.org/10.1111/j.1467-954X.2004.00459.x

Teerds, H. (2022). 'The space between': An architectural examination of Hannah Arendt's notions of 'public space' and 'world'. *The Journal of Architecture*, *27*(5–6), 1–21. https://doi.org/10.1080/13602365.2022.2137557

United Nations (2021). *The sustainable development goals report 2021: Goal 11 – Make cities and human settlements inclusive, safe, resilient and sustainable*. United Nations. Available at: https://unstats.un.org/sdgs/report/2021/Goal-11/ (Accessed: 16 April 2025).

Villa, D. R. (1996). *Arendt and Heidegger: The fate of the political*. Princeton, NJ: Princeton University Press.

Voice, P. (2013). Consuming the world: Hannah Arendt on politics and the environment. *Journal of International Political Theory*, *9*(2), 178–193. https://doi.org/10.3366/jipt.2013.0054

Ward, P. M. (2019). Self-help housing. In *The Wiley Blackwell Encyclopedia of Urban and Regional Studies*, edited by A. Orum. Hoboken, NJ: Wiley Blackwell. https://doi.org/10.1002/9781118568446.eurs0281

Zerilli, L. (2005). "We feel our freedom": Imagination and judgment in the thought of Hannah Arendt. *Political Theory*, *33*(2), 155–188. https://doi.org/10.1177/0090591704272958

10 Sensing the "Public" in Public Wall Art in Bangalore

Salila Vanka

Introduction

This chapter examines recent collaborative public wall art projects involving state actors, artists, and citizens in Bangalore, viewing public art as a "practice that constitutes a public by engaging people in political debate" (Deutsche, 1992: 39). It explores the extent and ways in which prevalent state-sanctioned public art projects in Bangalore engage with issues of social and spatial justice by emphasising "the fair and equitable distribution in space of socially valued resources and the opportunities to use them" (Soja, 2009: 2). The South Indian city of Bangalore (officially renamed Bengaluru), forms the backdrop of this study which, in the post-1990s economic reforms period, like other major Indian cities, experienced rapid economic, demographic, and spatial transformations. The capital city of the South Indian state of Karnataka, Bangalore's journey from a Garden City in the 20th century to a leading Information Technology City today, marks the city's conversion to a bustling metropolis with an insatiable appetite for land and resources.

In this aspiring world-class city, political and governing bodies seek inspiration from non-Indian models of urban development and management. Thus, inward-looking gated residential communities, corporate office towers, malls, and theme parks, continue to proliferate, replacing existing built forms. In this increasingly homogenised *milieu*, where state and private actors' claims supersede every other claim to the city, "there is a deep sense of loss of a place[-]based identity that this city, and its once distinct neighbourhoods, seems to have given up" (Mistry, 2017: 152).

The public walls of this socially and spatially fragmented metropolis provided the canvas for the *Karnataka Vaibhava* (meaning "The Grandeur of Karnataka" in Karnataka's state language Kannada) wall beautification project, initiated in 2009 by the local municipal body *Bruhat Bengaluru Mahanagara Palike* (BBMP). BBMP officials commissioned wall paintings that include "scenes from mythology, paintings of the temples at Hampi, Belur and Halebid, the Mysore Palace, historical figures of Karnataka, jungle creatures, and even a few socially conscious paintings of polio vaccinations and the dangers of alcohol related abuse" (Keshav, 2010: np). Covering the city's flyovers, underpasses,

DOI: 10.4324/9781003335993-10

Figure 10.1 A painting of the erstwhile king of Mysore State in the backdrop of temple plinths on the walls of an underpass in Bangalore. Part of a series of city-wide public wall paintings that the city authorities commissioned in 2009 (Source: Salila Vanka).

public buildings, and compound walls, the *Karnataka Vaibhava* project paintings had the two-pronged intent of discouraging the vandalism of public property and celebrating Karnataka's culture. The paintings drew upon imagery ranging from historic temples and palaces of the previous rulers of Mysore State (which included Karnataka) alongside civic messages (Figure 10.1).

Where globalising forces tend to homogenise and standardise the built environment, Bangalore's municipal authorities employed wall art to draw the viewer's eye to the specificity of the local as a spectacle for the local/global citizen's consumption. The authorities also sought to distract the viewer's eye from the ubiquitous litter, encroachments, and chaos that defy all efforts to organise and discipline public space and public conduct. Arguably, the wall paintings also represented state actors' efforts to "rule by aesthetics" which "is a process of translating broad aesthetic codes into a governing lens for organizing urban space" (Ghertner, 2011: 127).

Describing space as a social construct, Lefebvre states there is a politics of space because space is political (Elden, 2007). By the same token, where there are people, there is the possibility for politics. In a democratic state, urban public space offers a staging ground for contestations, conversations, and negotiations between diverse groups and actors of state and society (Deutsche, 1992; Harvey, 2008). With reference to the urban milieu, "the presence of 'the aesthetic'—whether embodied in artworks, architectural style, urban design, or museums—helps give redevelopment democratic legitimacy since, like the public, art often connotes universality, openness, inclusion" (Deutsche, 1992: 37). The public walls of the city might then provide blank canvases where public art has the agency to reflect urban inequities and injustices (Landau-Donnelly, 2023).

In Indian cities, the public realm

> shows a dissonance at many levels between two ways of dealing with and acting upon the world—of the poor and the rich, of the powerful and the dominated, of those who own property and those who can only defile it, of legality and illegality, of obedience and furtive evasion of rules.
>
> (Kaviraj, 1997: 84)

Kaviraj, here, was writing about Kolkata (formerly known as Calcutta), but the description is apt for Bangalore too, where societal inequalities are in full public view as affluent gated residential communities share boundaries with illegal squatter settlements, and informal street economies jostle for road space with buildings and vehicles serving the formal sector.

Against this backdrop, this chapter examines three collaborative public wall art projects involving state actors, local artists, design collectives, and citizens that were undertaken between 2019 and 2022. This research adopted visual analysis and content analysis of the three case studies. The study was conducted during the COVID-19 pandemic in 2021 and therefore relied on digital approaches such as public Zoom discussions, YouTube videos, and media coverage on the three cases. Other data sources included interviews with artists (who were not involved in the case studies), field visits, and photo-documentation at the project sites.

Inquiring into whether collaborative public wall art projects in Bangalore engage with issues of social and spatial justice, this research concludes that such projects represent the "aesthetics of governmentality" (Ghertner, 2011) and are not political in intent or content. However, this study also shows that collaborative art practices involving multiple actors of state and society represent the city as an *oeuvre*—"a collective artwork of all inhabitants and users in a city" (Lefebvre in Boer and De Vries, 2009: 1322). In the process of collectively making art, the projects might nonetheless help to create an opening from which the space for the politics of difference and diversity in Bangalore's public realm to occur in the future.

On the Politics of Public Art, Aesthetics, and Public Space in Indian Cities

Public art is a manifestation of urban aesthetics, where "aesthetic" designates, "among other things, a kind of object, a kind of judgment, a kind of attitude,

[...] experience [...or] value" (Shelley, 2022: np). This research departs from the premise that public art

> can affect how people perceive places, whether they feel like they belong (or not), whether they matter, feel safe, are welcome (or not)—ranging from interventions into existing, concrete systems of political power to broadening the very notion of what and where 'politics' is.
>
> (Landau-Donnelly, 2023: 28)

In its most universal definition, public art refers to a work of art in a public place that occupies space in the public realm. Beyond its material form, "the plasticity, materiality and hapticity of artistic media—including but not limited to artefacts, dance, installations, interventions, sculpture, music, painting, performance, photographs, smellscapes, soundscapes—are utterly connected with notions of space, place and placemaking" (Landau-Donnelly, 2023: 27). Arguably, these connections make public art an indicator of the extent to which citizens are engaging in the (non-)public realm.

Street art and graffiti are prominent forms of public wall art that represent "radical aesthetic practices, as rebellious, recalcitrant, even revolutionary acts" (Schacter, 2014: 161). Evolving in the 1970s in Europe and the United States, graffiti is a "primary form of 'deviant' aesthetic intervention in urban space" (Young, 2012: 2). Street art grew in the 1990s and is part of a counter-culture opposing rules (Ulmer, 2017). Graffiti artists use aerosol cans and marker pens and add their own unique tags to artwork on walls, "marking [...] territories and surfaces" (Halsey & Young, 2002, p.180). Street art uses varied media including "stencils [...] street sculpture [...], stickers, 'paste-ups' or 'wheat-pastes' (in which posters made by the artist are glued onto [...] surfaces), alteration of the words or image on a billboard, or performance-based actions" (Young, 2012: 2).

Depending on their content, graffiti and street art push the limits of legality and can be subject to censure and legal penalties. Public art that is driven by a "desire to alter the environment" (Young, 2012: 3), can be a gauge of tolerance to dissent and difference (Deutsche, 1992). Public art can also be political, as "artistic practices play a role in the constitution and maintenance of a given symbolic order or in its challenging and this is why they necessarily have a political dimension" (Mouffe, 2008: 4). Rancière (2011: 12–13) argues that "politics revolves around what is seen and what can be said about it, around who has the ability to see and the talent to speak, around the properties of spaces and the possibilities of time". In Indian cities, governmental and private actors commonly use wall art to address "the need for 'order' and 'discipline' in confronting the chaos of the city" (Benjamin, 2010: np). A common example is the practice of placing images of Indian gods on public walls to deter urination and spitting (Figure 10.2).

In Indian cities, painting on a public wall without governmental permission is illegal. What the artist considers freedom of expression in the public realm is

Figure 10.2 Signs and tiles with messages and images of Indian gods on a public wall as deterrents to vandalism and littering (Source: Salila Vanka).

punishable as defacement of public property. In addition to the warning signs around the fines for violations, the plethora of wall paintings signifies that the space around the walls is being maintained (by either private or public authorities) and is thus off limits to uncivil conduct. In this backdrop, more often than not, state-sanctioned street art in Indian cities manifests "as an art form almost entirely bound to a corporate, institutional visual regime" (Schacter, 2014: 162). This raises the question of whether state-sanctioned public art can be political (i.e. challenging the status quo in all aspects of urban life), and to answer this, it is necessary to first understand the nature of the public realm in Indian cities.

The "use of the term 'public' to describe a type of urban space, one accessible to all of a town's residents and owned by none in particular, probably began in medieval Europe" (Glover, 2008: 211). Governmental regulations of public space in post-colonial Indian cities originate in Eurocentric spatial constructs that differ from pre-colonial formations and uses of public and private space (Glover, 2008, Kaviraj, 1997, Chakrabarty, 1991). Public spaces in Indian cities are under state jurisdiction and are subject to appropriations and encroachments for different formal and informal activities.

Commenting on public spatial practices in post-Independence Calcutta, Kaviraj (1997: 84) describes municipal signages enforcing public cleanliness as a colonial legacy which "posited a duality between the city and country in which the city was seen as orderly, hygienic, scientific, technologically superior, and 'civilized'". This civilised, modern city depended on "civic consciousness and public health, of even certain ideas of beauty related to the management of public space and interests, an order of aesthetics from which the ideals of public health and hygiene cannot be separated" (Chakrabarty, 1991: 16).

In contrast, public space in pre-colonial India was regarded as a kind of "outside" space, "the space for which one did not have responsibility, or which was not one's own, and it therefore lacked any association with obligation, because it did not symbolise any significant principle [or] express any values" (Kaviraj, 1997: 98). Pre-colonial public space, then, was not about a controlled, civilised, or regulated realm, but the spaces that were not. This separation and tension between inside and outside, regulated and unregulated, private and public, civilised and uncivilised space continues to manifest in littering, vandalism, encroachments, and violence in the public realm in Indian cities. Destruction of public property such as government buses and street furniture are common expressions of collective protests against unpopular governmental actions.

In a milieu where state actors pursue unified narratives of urbanism in a top-down manner, "the production of aesthetics remain[s] disciplined within the rationale of the plan, and in effect, [... is the] spatialization of a particular power structure" (Benjamin, 2010: 2). State regulations and policing of the public realm transform the city into a spectacle where "it can then only be consumed passively, rather than actively created by the populace at large through political participation" (Harvey, 2008: 22). What, then, is the nature and role of public art in a city designed as spectacle?

On Public Art and the Public Realm in post-1990s Bangalore

In the opening chapter of his book "Art as Experience", Dewey proposes:

> to understand the aesthetic [...] one must begin with it in the raw; in the events and scenes that hold the attentive eye and ear of man, arousing his [sic] interest and affording him enjoyment as he looks and listens: the sights that hold the crowd—the fire-engine rushing by; the machines excavating enormous holes in the earth; the human-fly climbing the steeple-side; the men perched high in air on girders, throwing and catching red-hot bolts.
>
> (Dewey, 1958: 5)

Following Dewey, it is important to ask—what raw material do public spaces in Bangalore provide its artists? In an overloaded city with inadequate public services and ever-growing informal activities, public spaces such as streets, sidewalks, parks, and playgrounds are receptacles of the daily detritus of human and non-human activities. In their struggles for public resources,

> politicised linguistic, caste or ethnic groups frequently frustrate the ideologies of (corporate) beauty by expressing their 'desire for democracy', rather than modernity, thereby disrupting the rational visions of the town planners and citizens who cherish the image of a city that will take its proud place in a global capitalist order.
>
> (Nair, 2005: 297)

This "desire for democracy" is also evident in the plethora of graffiti and wall art across the city that communicate in graphics and words with passers-by. Movie posters, graffiti, and stains from spitting and urination compete with disciplinary messages such as "do not urinate" and "littering is punishable" on the city's public walls. The role of state-sanctioned public art in this milieu is primarily to deter acts of vandalism and littering that render the public realm unsightly, unusable, and unsafe in the eyes of officials.

Maintaining cleanliness in public spaces is a governmental function and in Bangalore (as in other Indian cities), sweepers and waste pickers, known as *pourakarmikas* in Kannada, are employed by BBMP to manually clean the streets and other public spaces (Figure 10.3). Wearing BBMP coats and armed with brooms, baskets, and wheelbarrows, the *pourakarmikas* sweep and pick

Figure 10.3 *Pourakarmikas* wearing BBMP coats at work on Bangalore streets with brooms and trash bags. The sidewalk has been appropriated by a builder to store construction materials for a building project (Source: Salila Vanka).

waste from their assigned neighbourhoods each morning. Street-cleaning does not conventionally fit in the framework of urban aesthetics, but arguably a litter-free city is a visually pleasing city and the act of cleaning is the act of beautification itself. At the same time, the common household act of making rangoli—where colourful and detailed patterns are created on the ground)—also contributes to the city's *oeuvre*, where residents become the artists who are shaping the "beautified" public realm. One might ask, then, whether the rangoli makers are themselves trying to reclaim their pieces of the public realm from those who might be seen to deface it and how does one draw the line between "beautiful" public art and defacement?

Bangalore had a state-appointed Urban Arts Commission—of ex-bureaucrats, artists, and architects—whose role was to provide a "visual aesthetic framework" (Kappan, 2020: np) for the city's built environment, but it went defunct in 2002. In the absence of a citywide vision for the city, the BBMP stepped in to beautify the city. In 2009, the BBMP initiated the Karnataka Vaibhava wall painting project, its purpose being "to avoid posters, wall writings, to bring about beautification, and to show the heritage and bring back the glory of Karnataka" (Keshav, 2010: np). Around the same period, there was a shift from handmade publicity posters for movies and political events to digital printing which was easily available and affordable. In interviews with the author, veteran Bangalore-based artists Ravi Kashi and Pradeep Kambathalli observed that change to digital printing caused artists and painters of film posters and political banners to lose their jobs. The BBMP initially reached out to Bangalore's art schools to undertake the wall art, but ultimately commissioned the unemployed painters who were willing to work at cheaper rates (Keshav, 2010).

In 2016, the BBMP Commissioner initiated a plan to repaint the city walls at the behest of the State Government. In his note to the finance department officials, the Commissioner wrote:

> The Kannada Development Authority (KDA) president has written to us, asking BBMP to take up painting work in Bengaluru city. It says that by painting the walls, we can maintain cleanliness near compound walls. BBMP has done the same in the past and by adopting the same method again, will also help promote tourism and culture of the state
>
> (Chaturvedi, 2016: np)

The BBMP wall painting project suggests that the local government saw aestheticisation as a means to maintain cleanliness in the city's public spaces. Around the same timeframe, there emerged a parallel culture of collaborative public artwork by artists who worked "with meager resources, often engaging with site-specific mediations outside of the gallery space" (Sawant, 2012: 126). A full discussion of Bangalore-based artists such as John Devaraj and Pushpamala N. and their practices outside the studio is beyond the scope of this chapter, but they serve as a good example of this trend. Noteworthy in their works, is a "shift from object-based works to dematerialized art practice", prioritising the idea behind the art over its physical form (Sawant, 2012: 125). These "dematerialised art practices" (some of them with public involvement) include installations in the city's deteriorating public resources—dying lakes, abandoned wells, and squatter housing to "remind the thronging crowds of the many contradictions in the neo-liberal agenda [...] altering the city's landscape and pushing it towards unsustainable, overindulgent consumption" (Sawant, 2012: 128).

Shortly after the wave of the BBMP wall paintings, came the Bangalore Street Art Festival in 2016, which included contributions from a leading school of art—the Srishti School of Art, Design and Technology (now known as SMI)—along with individuals and artist collectives from in and outside India, to create high-quality paintings in prominent public spaces. These works adorn the exteriors and interiors of the city's Metro rail stations, a historic market, and public walls (Figures 10.4 and 10.5). These projects had the approval and partnership of government authorities. The themes did not always attempt to be context-specific, but the colourful wall art attracted attention because of their prominent locations and eclectic content.

Notably, 2016 also marked the birth of Aravani Art Society, a "trans-women and cis-women led art collective that aims to create a space for people from the transgender community, to come together to create, collaborate and work on several artistic projects" (Aravani Art Society, 2021). The Aravani collective

Figure 10.4 Artist Shilo Shiv Suleman's mural on the lift shaft of Bangalore's oldest wholesale market, Krishna Rajendra Market, as part of the ST+ART India Art Festival in October 2016 (Source: Salila Vanka).

Figure 10.5 Mural by Indian artist Harshvardhan Kadam and students of Srishti Institute of Art, Design and Technology on the Church Street Metro Station entrance wall, as part of the ST+ART India Art Festival in October 2016 (Source: Salila Vanka).

has been active in getting art commissions from public and private organisations across the country.

To avoid interpreting all collaborative art practices as "compromised arrangements", Sawant (2012: 125) suggests it is

> crucial to distinguish between art practices that use collaboration to maintain the status quo through a series of invisible, coercive mechanisms and those that genuinely seek to make a difference by mobilizing social forces for meaningful interventions in the ways we think.

These various instances of wall art projects in Bangalore reveal that there is a range of actors and mechanisms through which art is being made.

There are different degrees of state sanction in each of the wall art projects. The degrees of freedom of expression vary with the amount of governmental presence. The state-sanctioned paintings had no participatory component. The dematerialised practices had citizen engagement but their ephemeral nature and small scale restrict their outreach to chance passersby or a niche audience. Against this backdrop, the collaborative projects I now turn to represent a new format of public art practice in Bangalore which are neither entirely top-down nor entirely grassroots-driven.

Three Cases of Collaborative Public Wall Art in Bangalore

Lum (2014: np) challenges how public art is often seen as a public "good", asking

> [w]hose interest does public art serve? Is it enough for a public artwork to be intellectually interesting, aesthetically pleasing, or to add to the character of a city? Or, in assessing the value of public art, should we consider the public 'good' in a broader context?

These questions resonate with this research which seeks to understand if collaborative art practices have the potential to expand spaces of engagement between different city publics through their public art work.

To explore these questions, I now examine three recent collaborative art projects in three different city locations. The three projects are: (i) "Essential" on Public Utility Building, Mahatma Gandhi Road, (ii) Richards Town Children's Wall Art Project, Richards Town in Bangalore Cantonment, and (iii) Malleswaram *Hogona* (Let's go to Malleswaram) in Malleswaram, a historic suburb. The "Essential" mural of a *pourakarmika* and informal waste collector celebrates the contribution of waste pickers during the COVID-19 pandemic. In the Richards Town Children's Wall Art Project, the neighbourhood residents used art to discourage vandalism and littering at a public wall. Malleswaram *Hogona* is a collaborative wall art project aimed to enhance walkability and imageability in hitherto hidden lanes of Malleswaram.

These three art projects were created between 2019 and 2021 and occupy prominent public walls and buildings and are apt case studies for the following reasons. First, they are collaborations between artist collectives, government institutions, non-governmental institutions, and members of socially marginalised groups. Second, they highlight social issues or cultural references that are specific to residents of Bangalore. Third, the artworks highlight important issues or groups, giving them visibility and recognition, while also acknowledging the significance of their creators or subjects. Fourth, they are in prominent locations of the city, thus laying claim to their place (and therefore their subjects' place) in Bangalore's public realm and discourses of social justice in the city. Fifth, all three cases are related to issues of garbage/littering and celebrate the *pourakarmikas* who work to clean the city's public spaces.

Essential (2021)

The "Essential" mural was painted on the entrance wall to the iconic Public Utility Building, a 55-year-old government building which stands on Mahatma Gandhi Road (MG Road), one of Bangalore's busiest streets since its construction during colonial rule (Figure 10.6). The mural is a collaborative effort between the BBMP, Fearless Collective (an art-activist organisation), and

Figure 10.6 "Essential" is a mural on the entrance block of the Public Utility Building, MG Road, Bangalore and represents a collaborative effort between Fearless Collective, BBMP, and Hasiru Dala (Source: Salila Vanka).

Hasiru Dala—a social impact organisation working for the rights of waste pickers, *pourakarmikas*, and volunteers. Painted during the COVID-19 pandemic, the mural celebrates the essential but unnoticed work of *pourakarmikas* in cleaning the city's streets at a time when the city was in lockdown and residents stayed safely indoors.

From the late 1700s to the 1990s, MG Road (known as South Parade during the British rule) was the main spine of Bangalore's central business district, with numerous outlets for retail, food, and recreation. As the city expanded, other business districts have emerged but MG Road still remains a busy transit corridor and a popular destination for local and outside visitors. Thus, the location of "Essential" on the tallest building of the most prominent street in the city centre is noteworthy, given its otherwise uncelebrated subjects.

In a video titled "We Are For You and You Are For Us" on the making of the "'Essential' project in August 2022 members of Fearless Collective describe it as "a public monument affirming the absolutely essential role women waste-pickers and *pourakarmikas* play in regenerating our cities and mitigating climate change" (Fearless Collective, 2022). In the video, *pourakarmikas* joined the artists, *Hasiru Dala* (meaning "Green Crusaders"), and the BBMP to discuss the mural, discussing their work and aspirations for city life which are rooted in sustenance and reciprocity. The video describes how the mural connects their indispensable work of cleaning the city for all city dwellers with what is important for the *pourakarmikas*—their need for dignity, respect, and recognition at their work—with the ephemeral *kōlam* (Figure 10.7) which is seen as beautiful and life-sustaining. *Kōlam*, similar to *rangoli*, is a traditional form of Indian floor art but differs in regional context, style, and specific cultural meanings. In drawing this connection between *kōlam* and the *pourakarmikas*, it seeks to impart the deep cultural, spiritual, and social significance of *kōlam* onto the *pourakarmikas* whose work is often undervalued but is vital. As the video describes:

> Early in the morning, after cleaning the ground at the thresholds of the entrances of their homes, women lay *kōlam* patterns on the earth, of rice and lime. These intricate geometries are an everyday reminder of beauty, inviting abundance and prosperity into our homes. But the purpose of *kōlam* is more than decoration. Traditionally, laying rice at the entrance of one's home was an offering to feed insects and ants [...] Women who work in Bangalore's waste management sector, chose *kōlam* as a symbol to celebrate the work that they do every day to regenerate the city. [...] These women act as a regenerative force. On a landmark building in the city, Jayabai, an informal waste-picker, gathers things that people consider wasted—bottles caps and wrappers, and gives them a new life. She embraces *pourakarmika* Valli—who turns organic matter—bones and love letters, back into the earth.

Figure 10.7 A *kōlam* at the doorstep of a household in the locality of Ulsoor in Bangalore. The *kōlam* is a ritual design traditionally made of rice flour that women in Hindu households make at the threshold of their home after washing the space every morning (Source: Salila Vanka).

In the same video, *Hasiru Dala*'s co-founder, Nalini Sekhar described the *pourakarmikas'* response to being acknowledged in the mural:

> Waste-pickers didn't even see themselves as contributors to the economy, to the environment. [...] So, that pride that needed to come, we really worked hard on that, and showed that they are actually saving Bangalore Municipality 84 crore rupees with their work. Once they got to know about it, they started talking about their work with a lot of pride.

The waste pickers' role in maintaining the city came into spotlight during the COVID-19 pandemic when they continued their daily jobs, risking exposure to COVID-19-related waste, since the BBMP did not provide all of them with personal protective equipment. Even as the "Essential" mural was being publicised as a tribute to the city's "frontline" workers (as the *pourakarmikas* were called), the *pourakarmikas* were holding protests at the BBMP headquarters demanding better salaries and working conditions.

The "Essential" project is distinctive for its theme, location, and intent—it is a collaborative art project that celebrates the humble waste pickers and their irreplaceable contribution to making the city clean and habitable. Cleanliness and beautification are recurring themes in state-driven discourses on the development and management of Indian cities (Baviskar, 2003, Nair, 2005, Benjamin, 2008). Regardless of the overburdened, underfunded, and mismanaged municipal services, the city's informal economic sector and squatter settlements are considered as major generators of garbage and dirt that interfere with this "socially produced aesthetic [...] that lays down boundaries between the beautiful/ugly, visible/invisible, legal/illegal" (Ghertner, 2011: 281). Against this backdrop, the acknowledgement of Bangalore's waste pickers in art (and as art), however symbolic, is a gesture of respect and an empowering moment for the *pourakarmikas*. The everyday act of cleaning and the *kōlam* as a motif of beauty and abundance—these are acts of creating an *oeuvre* in the city's public realm, with the *pourakarmikas* as the artists and the muse.

Richards Town Children's Wall Art Project (2019)

The Richards Town Children's Wall Art project is a mural representing children's perceptions of their neighbourhood on a public wall, around which had become the site for littering. This case highlights a local resident group's effort to curb littering and vandalism in a public space by using wall art. Located in Richards Town locality in the Bangalore Cantonment area, the project involved the Richards Town Residents' Association (RTRA), the Indian Railways, artists, traffic police, *pourakarmikas*, and children of the residents and *pourakarmikas*. The author was unable to take pictures because the wall was demolished before August 2023. An article titled "This wall has become the talk of the town" (Barkha K. 2019) in a local daily Bangalore Mirror features an image of the mural.

The artwork was painted on a wall on Indian Railways property at the edge of Richards Town, facing Clarence High School. Describing the area as a "huge garbage dump" prior to the project, an RTRA member said (in an online event via Zoom) that the India Foundation for the Arts (IFA) conducted in October 2020):

> There were animal carcasses, beer bottles, plastics, broken sanitaryware, broken furniture, the works. And they had the problem of flashers coming and harassing school children [...] So we decided to take up this spot,

> the broken-down wall, approach the railways and get it rebuilt, and clear it up. And after four months of relentless effort, the railways did build us a wall. But there were a lot of girders there and loose hanging cables, so we sought the assistance of the BBMP stakeholders. They came forward, they cleared the girders, they compacted the pavement with fresh soil, removed all the loose cables, and they fixed mesh over and above the wall to deter people from throwing garbage over the wall.

BBMP fixed the wall, but RTRA found the organisation's response to the garbage issue to be "at best, sluggish" (Pinto, 2020). The residents decided to take matters into their own hands and approached a local artist and the IFA, a Bangalore-based non-governmental organisation that supports projects and research in arts, for artistic and financial support, respectively.

Following IFA's mandate that the project should be inclusive, the RTRA members discussed what it meant to be part of a community. Pinto (2020: np) elaborated,

> We did some soul-searching... What does it mean to be inclusive? Does Richards Town belong only to the people living in the apartments? What about the PKs [*pourakarmikas*] who are so anonymous? They come early morning; before you wake up, they are gone. Does it not belong to them? They spend most of their [working] day [there]. What about the children of Clarence School? That's why we decided that it was to be truly and authentically inclusive, that we must recognise the contributors. The town belongs to so many others apart from those of us who stay in the apartments.

This recognition that Richards Town belonged to multiple groups motivated the RTRA members to initiate the wall art project as a collaboration of the residents, the school, and *pourakarmikas*, with children as the main artists. In a day-long event, the RTRA members, *pourakarmikas*, and children of the RTRA members, *pourakarmikas*, and Clarence High School, painted 107 feet of the wall. In the weeks leading to the day of painting, the artist walked with the children around Richards Town to sketch their impressions of the locality. The wall art was composed of the children's sketches and *kōlam* by the *pourakarmikas*. As the painting went on, the local traffic policemen helped in keeping the area safe from passing vehicles and even participated in the wall painting. The RTRA was also able to get a paint business to donate buckets of paint for the artwork.

The Richards Town wall art project is an illustration of a citizen-driven initiative that used art to involve community members and local government to address a civic issue. However, it also highlights a recurring aspect of public space management in Indian cities as an economic class issue, where tax-paying, propertied middle and upper-class groups claim greater rights to the city than their less-privileged counterparts living in illegal squatter settlements and working in the informal economy. Since the 1990s, Indian urban middle-class groups are seen to have gained greater access to the offices of local governing bodies due to new laws that mandated participatory governance (Fernandes, 2004; Kamath and Vijayabaskar, 2009). In this milieu, "upper class concerns around aesthetics, leisure, safety, and health have come to significantly shape the disposition of urban spaces" (Baviskar, 2003: 90).

In Bangalore too, vocal middle-class groups have been able to partner with state actors in shaping and managing public parks, lakes, and sidewalks. They have done so to not only discourage disorderly behaviour (which includes playing, eating, exercising, and bringing pets and toys) but also to restrict access to people who are not like them. The appropriation and regulation of the city's public spaces to keep them clean and beautiful might then restrict their role as sites where difference and diversity find ways to co-exist. The Richards Town wall art project, however, is a case where numerous classes, including the middle-classes, were involved and intervened into the city's public spaces also revealing the possibilities that collaborative art practices can present (even momentarily) to bringing together diverse social groups in addressing common concerns.

Malleswaram Hogona (2021)

The Malleswaram *Hogona* ("Let's go to Malleswaram") street art project is a collaborative public art initiative that was planned and executed during the COVID-19 lockdown in Bangalore. In early 2021, a group comprising Purpose Lab (a social impact organisation that started a mobility initiative called "Bengaluru Moving"), Malleswaram Social (a community group in Malleswaram), and *Geechegalu* (an artists' collective), worked together with the BBMP and other partners to conceptualise a walking tour through the streets of this historic suburb of Bangalore. Covering a distance of 1.5 kilometres, the project aimed to enhance the walkability and imageability of Malleswaram, an early 20th-century suburb built in the aftermath of a plague in Bangalore in 1898. In the years that followed, the city authorities built new suburbs to decongest the original settlement, mixing principles of modern planning with indigenous cultural practices rooted in social hierarchies.

An upper-middle-class mixed-use locality today, Malleswaram's residential areas are characterised by spacious properties and conservancy lanes. While serving as a separator between two properties to bring in light and ventilation, the conservancy lane was provided to facilitate the manual collection of human excreta from households. With the advent of modern services (and laws), this method of waste collection was discontinued many years ago. But the conservancy lanes remain as visual and spatial reminders of a socially unjust practice that marginalised people based on their caste.

In the days preceding the painting, the 13 artists involved in the project walked around the precinct and selected subjects that they felt best represented Malleswaram. The locations of the paintings were public and private walls (with the permission of the property owners) on the main streets and conservancy lanes which were different stages of use and cleanliness. The themes for the wall paintings included sparrows (a vanishing breed of birds from Indian cities), *sampige* flowers (a native Indian flower with the botanical name "Magnolia champaca"), a saree-clad woman's foot in mid-stride, an elderly woman pouring coffee in a steel tumbler (Bangalore is famous for filter coffee), a young woman standing amidst showers of marigold flowers which are used in celebratory events across India (Figure 10.8) welcoming visitors with folded hands, and a local *pourakarmika* (Figure 10.9), amongst others.

In a series of publicity videos shared by Bengaluru Moving in 2021, the artists of the Malleswaram *Hogona* project shared their choice of themes, experiences of working in the streets of Malleswaram, and local residents' responses to the paintings. Talking about the benefits of working with the government on street art, an artist said, "my earlier experiences with authority haven't been so good because street art has always been associated with vandalism. But here they coined it as beautification which is a pretty good start for a future that we want to see". In this context, the artist's remark highlights how rebranding street art as "beautification"—especially when linked to a social cause like

Figure 10.8 "*Svagata*" (meaning "welcome") is a painting by muralist Girija Hariharan subverting the discriminatory nature of conservancy lanes with images of a girl surrounded by marigold flowers and welcoming passersby with folded palms (Source: Salila Vanka).

Figure 10.9 Painting of *pourakarmika* Narayanamma who works in Malleswaram by artist Parameshwar Waran, who likens her quiet presence and impact on the city to that of the butterflies flitting around the locality (Source: Salila Vanka).

promoting walking—opens up opportunities for collaboration between artists and authorities. This shift in framing shows how collaborative art practices can challenge conventional views and create space for meaningful social engagement.

Another artist commented on the relation between street art and the material space it occupies, saying that street art is “less of your mark. The space owns the piece more than you own the piece. This is the first project where a city as a whole is interacting with me” (Varkey in Bengaluru Moving, 2021a: np). This highlights how street art, when placed in public spaces, is shaped by those spaces, becoming a shared experience between the artist and the city. Moreover, engaging in public art is seen as a great responsibility—as well as an opportunity, as these artworks interact with and influence the people who engage with the space. In an interview with a local daily newspaper, a Bangalore-based researcher highlighted the significance of the imagery—visual and social—by observing that,

> the process of redesigning a public space has to take into account a local sense of what is beautiful and what is acceptable to people who use or inhabit that particular space [...] One has to be conscious of whether the design invites rather than restricts use, movement and ownership.
>
> (Rao in Shree, 2023: np)

On the topic of repurposing the conservancy lanes as aesthetic spaces of public life, another artist said,

> when I learned about conservancy lanes, I could see a different type of people were helping these people live a very comfortable life, but they were entering through the backdoors [...] So I realised—let's have these marigolds coming out of the [drainage] pipe to show that there is beauty to be found anywhere, including conservancy lanes which were not technically considered the beautiful part of the city.
>
> (Hariharan in Bengaluru Moving, 2021b: np)

Another artist saw the opportunity to highlight the role of the *pourakarmika* as a contributor to the aesthetics of the street and maker of the city's *oeuvre*,

> I got a subject—a *pourakarmika*—one among those cleaning this street [...] many times, people asked me, 'Why her? Who is she?' I was very happy that there are questions raised from the public. Because no one knows what she works like.
>
> (Waran in Bengaluru Moving, 2021b: np)

The Malleswaram *Hogona* public art project was a collaborative placemaking and mobility initiative that highlighted the benefits of translating location-specific themes into public artwork that welcome engagement and participation by local communities. The project used public wall beautification to create walkable and inviting streets that encourage mobility among local residents. The project's aestheticisation of conservancy streets (which today are neglected receptacles of trash, or used as vehicular streets, or illegally privatised by adjoining property owners), to remake them as inclusive public spaces, also represents a gesture—however symbolic—to reimagine and remake the city's public realm. It underlines the idea of the city as an *oeuvre* that is made by the myriad everyday activities of its multiple publics.

Conclusions

Since the 2000s, there has been a proliferation of state-sanctioned public wall art in Bangalore with the two-pronged agenda of deterring vandalism in public spaces and beautifying the city. Bangalore's public walls were covered with images of local flora, fauna, landscapes, architectural monuments, and public personalities. In more recent times, there have been cases of state-sanctioned art where local artist collectives, state actors, design practices, and art institutions are collaborating to create public artwork that ostensibly taps into issues of social and spatial justice. This research posits that these art projects represent processes through "which one reimagines, remakes and re-experiences urban conditions and the notion of the urban itself" (Mistry, 2017: 151).

Solomon Benjamin (2010: 1) argues that the "relationship between aesthetics and politics is complex, hardly linear, involves city development processes that operate at separate levels and, most important, a multitude of actors". State actors use art as a means to clean and discipline public spaces and collective behaviour. Residents perceive art as a means to achieve neighbourhood beautification and civic pride. Artists see art as a means to create nostalgia, context, and social commentaries in the public realm, where the wall art projects are invitations to "change ourselves by changing the city" (Harvey, 2008: 23).

As a means to accept, and respond to, the diversity and differences in urban life, Solomon Benjamin (2010: 4) suggests that, "[r]ather than assume that cities must have a common aesthetics bound by a singular history or trajectory, we need to consider them as territories of multiple hues". Alongside disciplines such as architectural and environmental conservation, "public art can take on a special evocative role in helping to define a city's history if, and only if, they are complemented by a strong community process that establishes the context of social memory" (Hayden, 1995: 76). The three case studies in this research indicate that public artwork can be a common platform that brings together diverse groups in collective artistic representations of life in Bangalore.

The heaps of trash around many public walls in Bangalore suggest that it will take more than wall art to deter vandalism and littering. Moreover, it might be argued that state-sanctioned wall art projects present as "artwashing" (Schacter, 2014: 163) that seek to create the image of an inclusive city, rather than actually creating an inclusive city in terms of overcoming various social justice issues. An example of this form of "artwashing" would be recognising the *pourakarmikas*' work in art, with little regard to their working conditions that came to the fore during the pandemic.

This chapter has proposed a different lens that focuses on the myriad ways in which city residents amongst others participate in shaping the city as an *oeuvre* through their everyday social and spatial practices. In other words, the city is itself an artwork that is continuously formed and redefined by the people who inhabit and interact with it. The chapter has argued that, in Bangalore, state-supported collaborative public art practices do not *themselves* challenge the political status quo or raise issues of social and spatial justice. After all, they reflect the beautification vision of the BBMP and so serve as a kind of aesthetic governmentality, shaping what practices are welcome or unwelcome and serving as artwashing as the plight of the *pourakarmikas* endures (Ghertner, 2011).

However, this chapter also observed how collaborative public art practices involving state agencies, artists, activist groups, citizen volunteers, and social institutions, nonetheless create important spaces of engagement between the city and its people in multiple ways. First, they include diverse groups in dialogues and discussions about the shared public realm. Second, the diverse themes of the wall art represent the diversity of urban residents, experiences, and life in the city. Third, they give visibility and voices to invisible communities and celebrate their contributions to the city. Fourth, by aestheticising (or giving artistic form) to the un-aesthetic (garbage) or "others" (marginalised communities). Hence, despite the present lack of changes to matters of social and spatial justice, in transforming public walls into shared urban experiences, collaborative public art practices in Bangalore invite city dwellers to make a different sense of the city, and perhaps from the cracks that these activities create, a different politics of the city might emerge.

References

Aravani Art Society. (2021). Annual Report 2020–21. https://aravaniartproject.com/. June 15, 2023.

Barkha K. (2019). "This wall has become the talk of the town." *Bangalore Mirror*, November 04, 2019. https://bangaloremirror.indiatimes.com/bangalore/civic/this-wall-has-become-the-talk-of-the-town/articleshow/71881616.cms?utm_source=contentofinterest&utm_medium=text&utmcampaign=cppst (May 09, 2020)

Baviskar, A. (2003). "Between violence and desire: space, power, and identity in the making of Metropolitan Delhi". *International Social Science Journal*, 55(1), pp. 89–98. http://www.ingentaconnect.com/content/bpl/issj/2003/00000055/00000001/art00023 (February 02, 2023)

Benjamin, S. (2008). Occupancy urbanism: Radicalizing politics and economy beyond policy and programs. *International Journal of Urban and Regional Research*, 32(3), 719–729.

Benjamin, S. (2010). "The Aesthetics of 'the ground up' City: some insights from Bangalore." *Seminar*. https://www.academia.edu/111677168/The_aesthetics_of_the_ground_up_city_Some_insights_from_Bangalore. (August 7, 2024).

Bengaluru Moving. (2021a). Malleshwaram Hogona: Within her stride by Anpu Varkey [online]. Available from: https://www.youtube.com/watch?v=C9LMxfu1VoQ

Bengaluru Moving. (2021b). Malleshwaram Hogona: Svagata by Girija Hariharan [online]. Available from: https://www.youtube.com/watch?v=V3zbAd95reo&t=2s

Boer, R.W.J. and Vries, J. De. (2009). "The Right to the City as a Tool for Urban Social Movements: The Case of Barceloneta." In *The New Urban Question – Urbanism beyond Neo-Liberalism, The 4th International Conference of the International Forum on Urbanism (IFoU)*. Amsterdam/Delft. pp. 1321–1330.

Chakrabarty, D. (1991). "Open space/public place: Garbage, modernity and India." *South Asia: Journal of South Asian Studies*, 14(1). pp. 5–31.

Chaturvedi, A. (2016). "Walls in city to be painted again." *Bangalore Mirror*, March 28, 2016. https://bangaloremirror.indiatimes.com/bangalore/others/walls-in-city-to-be-painted-again/articleshowprint/51554477.cms?prtpage=1 (July 15, 2021).

Deutsche, Rosalyn. (1992). "Art and public space: Questions of democracy." *Social Text*, no. 33, pp. 34–53. http://www.jstor.org/stable/466433

Dewey, J. (1958). "Ch.1. The live creature." In *Art as Experience*. New York: Capricorn Books. pp. 3–19.

Elden, S. (2007). "There is a politics of space because space is political: Henri Lefebvre and the production of space." *Radical Philosophy Review*, 10(2), pp. 101–116.

Fearless Collective. (2022). https://www.youtube.com/watch?v=daLgOm8wzLc (October 02, 2022).

Fernandes, L. (2004). "The politics of forgetting: Class politics, state power and the restructuring of urban space in India." *Urban Studies*, 41(12), pp. 2415–2430.

Ghertner, D.A. (2011). "Ch.11. World-Class Detritus. *Rule by Aesthetics: World-Class City Making in Delhi*." In *Worlding Cities: Asian Experiments and the Art of Being Global*, edited by A. Roy and A. Ong. First Edition. Blackwell Publishing Ltd., pp. 279–306.

Glover, W.J. (2008). Construing urban space as "Public" in Colonial India: Some notes from the Punjab. *The Journal of Punjab Studies*, 15(1), pp. 1–14.

Halsey, M., & Young, A. (2002). The meanings of graffiti and municipal administration. *Australian & New Zealand Journal of Criminology*, 35(2), pp. 165–186.

Harvey, D. (2008). Right to the city. *New Left Review*, 53 (Sept–Oct) pp. 23–40.

Hayden, D. (1995). *The Power of Place: Urban Landscapes as Public History*. Cambridge, MA: The MIT Press.
India Foundation for the Arts. (2020). Throwback Thursday #11 | Art in the Neighbourhood: Exploring Richards Town | A Presentation by Anaheeta Pinto, October 22, 2020. https://indiaifa.org/events/throwback-thursday-11-art-neighbourhood-exploring-richards-town-presentation-anaheeta-pinto August 20, 2021.
Kamath, L. & Vijayabaskar, M. (2009). Limits and possibilities of middle class associations as urban collective actors. *Economic & Political Weekly*, 44 (26–27). pp. 368–376
Kaviraj, S. (1997). "Filth and the Public Sphere: Concepts and Practices of Space in Calcutta". In *Public Culture*. Durham, NC: Duke University Press, pp. 83–113
Kappan, R. (2020). "Can a revived urban arts commission restore Bengaluru's aesthetic beauty?" *Deccan Herald*, November 15, 2020. https://www.deccanherald.com/india/karnataka/bengaluru/can-a-revived-urban-arts-commission-restore-bengalurus-aesthetic-beauty-915690.html
Keshav, K. (2010). "Bouquets, brickbats for BBMP's wall paintings." *Citizen Matters*, February 23, 2010. https://bengaluru.citizenmatters.in/1798-bbmp-wall-painting-bengaluru-1798 (May 09, 2020)
Landau-Donnelly, F. (2023). "Art". In *Concise Encyclopedia of Human Geography* (Elgar Encyclopedias in the Social Sciences Series), edited by L. Lees and D. Demeritt, pp. 27–31
Lum. K. (2014). "Some reflections on urban public art today." *Penn Institute of Urban Research*. November 20, 2014. University of Pennsylvania. https://penniur.upenn.edu/publications/some-reflections-on-urban-public-art-today (August 30, 2022)
Mistry, A. (2017). "Art in transit: A pedagogic approach." *Visual Inquiry*, 6(2), pp. 149–164
Mouffe, C. (2008). "Art as an Agonistic Intervention in Public Space." In *Art as a Public Issue: How Art and its Institutions Reinvent the Public Dimension*, edited by Liesbeth Melis and Jorinde Seijdel. Rotterdam/Amsterdam NAI Publishers
Nair, J. (2005). "Ch.7. Battles for Bangalore: Reterritorializing the City." In *Promise of the Metropolis: Bangalore's Twentieth Century*. New Delhi: Oxford University Press, pp. 271–298
Rancière, J. (2011). "The Thinking of Dissensus. Politics and Aesthetics." In *Reading Rancière*, edited by P. Bowman and R. Stamp, New York: Continuum, pp. 1–17
Sawant, S. (2012). "Instituting artists' collectives: The Bangalore/ Bengaluru experiments with 'Solidarity economies'." *Journal of Transcultural Studies*, 3(1), pp. 122–149. https://doi.org/10.11588/ts.2012.1.9275
Schacter, R. (2014). "The ugly truth: Street art, graffiti and the creative city" *Art & the Public Sphere*, 3(2), pp. 161–176.
Shelley, J. (2022). "The Concept of the Aesthetic". In *Stanford Encyclopedia of Philosophy*, edited by E.N. Zalta. February 28, 2022. Available from: https://plato.stanford.edu/entries/aesthetic-concept/
Shree, D.N. (2023). Design to Fix Public Problems of Painted Walls and Clean Streets, January 28, 2023. https://www.deccanherald.com/india/karnataka/bengaluru/design-to-fix-public-problems-1185439.html
Soja, E. (2009). "The city and spatial justice." In *Paper Prepared for Presentation at the Conference Spatial Justice*, Nanterre, Paris, March 12–14, 2008
Ulmer, J. B. (2017). Writing urban space: Street art, democracy, and photographic cartography. *Cultural Studies↔ Critical Methodologies*, 17(6), pp. 491–502.
Young, A. (2012). Criminal images: The affective judgment of graffiti and street art. *Crime, Media, Culture*, 8(3), pp. 297–314.

11 Conclusion

What Next? Future Directions for Aesthetics and the City

Amy Barron and Joe Blakey

Introduction

This book began with the question, "how do we make sense of the city?". Such a question is relevant given that urban thought on the role and epistemological status of the city has arrived at somewhat of an impasse. On one hand, the city is seen as a presupposition, where "the city" imposes an unnecessary framework on how urban phenomena are understood. On the other, the city invariably still figures in the popular imagination, shaping, and being shaped by, everyday lived experiences, practices, and representations. This influence is apparent in the material and discursive artefacts that come from cities—speeches, books, and artworks—which can challenge, transform, or reinforce how cities, and the various urban ephemera they relate to, are perceived.

It is in aesthetics that this book has found a way to balance the caution against presuming too much about the city in advance and acknowledging the role of the city in shaping collective imaginations and individual lives. Through the chapters in this volume, we have demonstrated how it is through aesthetics that scholars can reflexively question how we understand—or should understand—the city. By highlighting our sensuous engagement with the world, aesthetics offers a perspective through which cities can be apprehended. This includes exploring the role of "common sense" around cities in shaping or reshaping urban spaces and experience, artistic and aesthetic practices about and "within" cities and, ultimately, imaginaries of a city.

These imaginaries, though, are far from singular, static, or self-contained. Rather, they are continually coming into being in-relation to and through representations, discourses, and lived experiences (Barron, 2021a, 2021b). Indeed, when we hear a city's name, particular images or practices might come to mind. Paris, for instance, might make us think of the "city of love", the Eiffel Tower, or Avenue des Champs-Élysées, whereas New York conjures images of skyscrapers, Central Park, and yellow taxis. These are common-sense reference points through which we come to understand and "know" cities that persist, affect, and are affected by urban experience most generally (Anderson, 2019). This book has highlighted that representations of cities do not exist *a priori*—and we should resist assuming them as such—but they are (re)imaged, performed,

DOI: 10.4324/9781003335993-11

and resisted in various ways, whether through everyday conversation, the reciting of a memory, or through the TV or newspaper. Common-sense ideas around what a city is, then, are far from singular, static, or self-contained.

In the midst of critiques of city-centric thinking, aesthetics stands out as a concept through which the city's relative importance in urban thought can be assessed and reaffirmed, whilst avoiding reification. While related concepts like discourse may address the power of representations or the apparent common-sense narratives surrounding cities, they tend to do so within specific theoretical and epistemological traditions, or at least a narrower theoretical aperture than aesthetics. In contrast, aesthetics offers a shared platform for scholars from diverse backgrounds to engage in critical conversations around and about the city. While some might take the different ways aesthetics is understood to mean it is somewhat of a blunted, vague, or ambivalent concept, this book has highlighted aesthetics instead as an opportunity to bring together scholars who share an interest in how we make sense of the world (Hawkins and Straughan, 2016). Through the chapters in this book, we have explored a range of aesthetic approaches within urban studies and have highlighted how one's interpretation of aesthetics deeply influences what can be understood about cities and the urban order more broadly (Murakami Wood and Abe, 2011). It is this blend—or perhaps even clash—of approaches that enables scholars to better understand how the city remains relevant to urban thought and questions.

We therefore encourage and stress what we would term as an "agonistic" approach to aesthetics, especially when approaching and debating the relative significance and role of the city or cities to urban thought (Mouffe, 2005). Agonism here refers to a constructive disagreement and dialogue between different positions. It is not about synthesising approaches to aesthetics, or aiming to add them together to create a whole like a jigsaw, but entails "a self-conscious dialogue with, criticism of, or opposition to, the intellectual, ethical or political limits" around alternative theoretical perspectives on aesthetics (Barry, Born, and Weszkalnys, 2008: 29). Doing so avoids any definitive theorisation of aesthetics, and by extension, the city, instead positioning them as subjects of ongoing reflexive—yet productive—disagreement and inquiry.

An agonistic approach to aesthetics therefore imparts and suggests a kind of "kaleidoscopic" understanding of the city, where radically different patterns pertaining to the city become knowable, depending on our theoretical and epistemological persuasion, and the grounded empirical particularities in terms of how the city is sensed or made sensible. As if seen through a kaleidoscope, the city becomes a complex and contingent construct, shifting with each turn of perspective, and at the same time, it implies there is no "master" perspective on either aesthetics or the city. Perhaps the limitation of such a metaphor, however, is that there is no guarantee that the city lies at the end of the kaleidoscope; for we should avoid presuming its relevance or qualities in advance. Either way, this approach encourages a meeting ground for difference, opening up possibilities for richer and reflexive engagements with the aesthetic, the city, and the urban, fostering an ongoing dialogue that resists closure or reductive consensus.

We then also caution that aesthetics is no silver bullet. Our intent in this volume has certainly not been to disregard those urban thinkers warning about reifying the city. Processes of urbanisation are certainly not reducible to some notion of the city, nor should "the city" be naturalised in any way, or defined in a manner that is absolute (Angelo and Wachsmuth, 2015; Brenner, 2019). If one is seeking to understand urbanisation more generally, then the warnings around methodological cityism—that taking "the city" as an entry point to understand urban processes leads to blind spots—are well-placed (Angelo and Wachsmuth, 2015). The task of this book, though, has been to think about the enduring role of the city and how we might attend to it. Our answer to the question of how we make sense of the city is not one that offers a particular definition of it. We have not sought to offer a model, heuristic, or theory "of" the city. Instead, our proposition is to suggest that we make sense of the city through aesthetics. If we have aesthetics at the forefront of our minds when considering the city, then it encourages critical questions around how we experience, perceive, and engage in them without distilling urban processes to them. To illustrate this, we now turn to how contributors used aesthetics to make sense of the city.

City Aesthetics

Aiming to avoid narrowing the conceptual scope of aesthetics and the city, the contributors were given the freedom to interpret both in a way that they chose. The intention was to leave open the possibility for radically different understandings and to draw attention to the contingencies shaping scholarly (and popular, in the case of the city) understandings of these notions. Contributors were asked, however, to define their approach to aesthetics and explain how it helped them to make sense of the city. While the chapters do this in different ways, and to varying degrees, our sense is that, collectively, they illustrate how aesthetics can be a vital concept through which we can collectively critically consider the city. We now work through their approaches in turn.

In Chapter 2, Sanna Lehtinen considered how aesthetics and the city have been understood through the discipline of philosophy, drawing on two layers of aesthetics: a "macro" layer, dealing with imagined, represented, and remembered elements, and a "micro" layer, addressing direct, sensory encounters with the city. Lehtinen advocates for a dual focus on everyday sensory perception and common ideas about cities, highlighting the importance of everyday life and the unplanned aspects of urban environments. Her approach emphasises the interdisciplinary nature of studying urban sensescapes, urging that it be a transdisciplinary effort.

In Chapter 3, Mónica Degen and Beatriz Gujarro discuss the central role of aesthetics in making sense of the city within the context of urban commodification, or the "aesthetic economy". Focusing on Barcelona and London, they explore how city branding is part of a broader aesthetic governmentality where authorities seek to shape a common-sense aesthetic by reconfiguring urban imaginaries through digital and analogue mediation. They show how sensory experiences are commodified to turn cities into products or ideas. The city then transcends traditional boundaries, becoming an idea, story, or discourse, while also being something to buy into or experience directly. The chapter highlights a need to consider both the lived and embodied dimensions of urban life (how discourses, representations, and imaginaries around the city work to order it) and how the city is variously encountered. They call attention to which understandings of the city are foregrounded and made visible, and which are resisted (Rancière, 1999).

In Chapter 4, Josh Widera returned to the original meaning of aesthetics as sensory perception to understand how bicycle messengers experience their work environments. Using ethnographic fieldnotes and ride-along interviews, Widera shows how their embodied experiences shape their sense of place and self, highlighting how their enjoyment elevates their experience beyond mere labour. By understanding aesthetics as a minor theory, Widera offers a non-exhaustive explanation of aesthetics, encouraging an integration of concepts including affect, corporeal experience, anaesthetics, and kinaesthetics. While anaesthetics shows how aesthetics can be a form of resilience, kinaesthetics situates the subject's knowledge within the city they are working in. By interpreting the city from an embodied disposition, we see how messengers produce a unique aesthetics of the city, building an intimate knowledge of the cities they navigate.

In Chapter 5, Friederike Landau-Donnelly and Stefanie Fredrik examined graffiti governance in Linz, Austria, as it strives to become the country's cleanest post-industrial city. They emphasise the link between cleanliness and dirt, highlighting the conflict between whose values of contamination and cleanliness shape the city. Introducing "infesthetics" (a portmanteau of "infest" and "aesthetics"), they describe graffiti as parasitic: as exploiting rather than serving the city. The authors mobilise a post-foundational view of contingency, considering what is (im)possible within particular social structures, noting that these parameters can be transformed through political acts. The authors argue for a conflict-aware aesthetics, showing how competing aesthetic visions exist within the city and how moments of conflict can be understood aesthetically (Marchart, 2019). Aesthetics, therefore, shapes what the city is, what belongs in it, and how these ideas are contested.

In Chapter 6, Günter Gassner critically and creatively addresses urban racism, borders, and imagination, proposing new perspectives against the violent imaginaries of the organismic city. He examines Nazi visions of a post-race city and how these relate to viewing the city as an organism. Gassner argues that urban imaginaries are contestable, yet some, like racist narratives, become dominant. Gassner contrasts the 1930s vision of a post-race city—an imaginary of a city free from experiential racisms, a celebration of racism that lacks racial outbreaks within its bounds (what he terms "racism without racisms")—with contemporary claims about the end of race, which overlook or deny structural racism while racisms nonetheless continue to be experientially felt in the city ("racisms without racism"). In doing so, he highlights the contingency of how cities are understood and the potential for political change by challenging hegemonic views. Gassner advocates for academics to combat these imaginaries through creative writing, offering alternative interpretations of the city. His approach aims to disrupt conventional academic forms and open new ways of making sense of the city and race.

In Chapter 7, Taher Abdel-Ghani, Yara Mohamed, and Amr Ibrahim used Deleuze's concept of the haptic to explore urban aesthetics as a tool for evoking cultural memory. They focus on graffiti at the East Side Gallery in Berlin and the Mohamed Mahmoud Wall in Cairo, examining how these urban canvases represent identity and history. Here, the city acts as the stakes for aesthetic encounters as its fabric is actively remade to narrate its past, present, and future. Adopting an approach attuned to movement and inspired by non-representational theories, the authors analyse how different art forms affect and are affected by the lives of those who inhabit Berlin and Cairo and the collective imaginaries of each city, rather than falling back on static representations. They introduce the concept of a mnemonic, where urban artworks serve as aids for retrieving and reproducing the collective memory of these cities.

In Chapter 8, Joe Blakey examines the parallels between the scale debate in human geography and the challenges to city-centric thinking in urban geography. He highlights how the city is often seen as a scale (the "city-scale") and notes that assuming fixed scales can hinder understandings of urban life "on the ground". Blakey suggests that we need not reject the city-scale. He draws on the aesthetic thought of Jacques Rancière who argues that life takes place amidst a sensory fabric called the "distribution of the sensible"—which polices everyday life and is the stakes of politics—to consider how the city scale figures in our common sense. The chapter highlights the importance of focusing on moments of dissensus regarding the city-scale to avoid imposing preconceived notions around the city into our studies. While acknowledging the significance of lived experiences, Blakey points out that certain power geometries, contingencies, and forms of common sense, like the city-scale, remain influential in shaping urban life as we know it.

In Chapter 9, Julian Brigstocke uses an Arendtian approach to aesthetics—emphasising imagination, the everyday, the provisional, and the micropolitical—to examine experimental urbanism projects at the Tijuana–San Diego border. He draws on Arendt's idea of a "sixth sense", which, while not tied to a specific sensory organ, creates a unique feeling of realness. Brigstocke argues that these cross-border urbanisms foster this sense of reality by embracing plurality, allowing individuals to understand the world from others' perspectives. He warns that a focus on city-centrism can overlook situated knowledge and everyday experiences where cities play a crucial role. Aesthetics, he suggests, not only contributes to debates about the city but also encourages exploration of how urban phenomena present themselves through distinct ways of sensing and inhabiting space. For Brigstocke, defining the city or the urban in rigid conceptual terms is less important than understanding how we make sense of the world through aesthetics. Understanding urban experiences through aesthetics avoids a fixed notion of the city, focusing instead on how urban experiences are shaped and understood.

In Chapter 10, Salila Vanka explores various thinkers on aesthetics, referencing Solomon Benjamin's idea that there is no singular aesthetic of the city. Vanka instead argues that cities should be viewed as "territories of multiple hues". Focusing on state-sanctioned public art in Bangalore, Vanka applies Ghertner's concept of aesthetic governmentality, where aesthetic norms of cleanliness, orderliness, and visual appeal serve as a means to govern urban space. These norms shape how urban life is experienced in Bangalore. While state-backed collaborative public art initiatives do not challenge the political status quo or address social and spatial justice, Vanka shows that they still create valuable spaces for resident engagement.

The chapters present interconnected arguments, with Lehtinen and Landau-Donnelley and Fredrik emphasising the need for a phenomenological focus on sensory experiences of the city, particularly through hands-on encounters related to *aisthēsis*. Widera, Abdel-Ghani, Mohamed, and Ibrahim highlight the importance of affect, movement, and lived experiences in understanding aesthetics. In adopting an approach to aesthetics attuned to affect and movement, these scholars prioritise the on-flow of urban sense-making, averting the trappings of any *a priori* notion of the city that might itself intrude upon how urban phenomena are understood. Both chapters explore how representations of the city are experienced: for Abdel-Ghani, Mohamed, and Ibrahim, it involves encounters that trigger shared and individual memories around the city, while for Widera, it relates to the imagined ideas of bicycle couriers shaped by their journeys and broader relationships (Anderson, 2019).

Other contributors highlighted the need to consider both sides of aesthetics: common-sense aesthetics rooted in representation and imagination, and *aisthēsis* which considers the lived, sensory encounters with the world. This dual approach is valuable for understanding the city, as it balances the risks of assuming fixed ideas about what the city "is" while recognising the shared imaginaries that shape everyday urban life. Such pieces emphasise the contingency of city imaginaries whilst avoiding naturalising them. By moving between common-sense aesthetics and *aisthēsis*, we can implement a kind of check and balance for considering the city without presuming what it is or overlooking blind spots in how urban life is lived. Indeed, several authors draw on post-foundational thinkers like Rancière and Arendt, who share this appreciation for the power of contingent shared imaginaries and the different ways they can be disrupted (Blakey et al., 2022; Landau-Donnelly and Pohl, 2023).

In other chapters, there is a consideration for the power of aesthetics in shaping urban space through city branding and the imposition of sensory qualities like cleanliness, orderliness, or visual appeal, which are often tied to a city's identity and which work to govern urban space. Asher Ghertner's (2015) work on aesthetic governmentality has been central here. Elsewhere, we note in Widera's chapter, a focus on how bicycle couriers develop resilience, contentment, and a sense of identity in the face of gig economy precarity and physical challenges. Similarly, Blakey's chapter also shows how individuals resist dominant ideas of the city, using political acts to challenge common-sense notions.

What Next for Aesthetics and the City?

With the above in mind, we propose the following six interrelated directions for future research on aesthetics and the city. First, whilst aesthetic thought has invariably attracted more creative methodologies (c.f. Arnold, 2019; Hawkins, 2020; Gandy, 2021), the potential of creative, mobile, and participatory methods for studying aesthetics and the city should be given more critical attention, without dismissing the value of traditional methods such as interviews (Hitchings, 2012). This book has not explicitly focused on method, yet Widera's ride-along interviews with bicycle couriers struck us as helpful in revealing the interplay between common-sense ideas of the city and how it is lived, whilst Gassner's use of journaling captured his own situated encounters with the two cities considered.

We detect that combining narrative-based, mobile, or visual methods—such as photo go-alongs or collages—offers further potential to illuminate how people make sense of themselves in relation to and through the city. Creative methods—those approaches associated with novelty, innovation, and perhaps even artistry or expression, and that go beyond those more rigid and established methods like interviews to tap into the messiness of the world (see Von Benzon et al., 2023)—strike us as particularly well-aligned with the nature of aesthetic inquiry. This is because they can facilitate an engagement with the more sensory and affective dimensions of experience. Creative approaches might enable a consideration of how participants—and the researcher—touch, move through, and experience phenomena, enabling a focus on the embodied nature of *aisthēsis*. Furthermore, they provide highly situated accounts of how specific common-sense understandings around the city are constructed, lived out, and perhaps even disrupted.

Second, and related, scholars might engage in creative ways of presenting their work, understanding this as an opportunity to actively intervene in how cities are understood. Whilst the written form arguably persists as the central mode of academic communication, it is increasingly being reshaped and expanded in scope by such creative writing practices (Hawkins, 2020). While all contributors explored both the power of common-sense understandings of the city and how these can be disrupted, we found Gassner's chapter and his use of creative writing particularly evocative. In the creative and iterative format deployed, oscillating between two journal-like entries between Bristol and Berlin, Gassner's writing form seeks to intervene in violent urban imaginaries. For Gassner, writing in this manner, which he terms as "non-organismic", becomes a way to contest "organismic" understandings of the city. In other words, resisting the hierarchical, centralised, and teleological characteristics typically associated with organisms through a writing form that works against these very characteristics, creating space for alternative ways of understanding the city.

Whether in moving towards post-race or post-colonial cities, or rendering visible and/or righting whatever injustice is carried out in, around, and in the name of "a city", the form scholarly work takes can play a role in challenging dominant narratives about cities—and this need not be limited to the written word (Hawkins, 2015, 2020, 2021). Of course, urban scholars are already engaged in various artistic practices—film, art, video, performance, podcasts, and blogs—that offer new ways to understand, address, make visible, or contest the various injustices surrounding urban life, and we encourage scholars interested in aesthetics and the city to harness these creative approaches to reimagine the city and what it stands for anew (Mason and Riding, 2023). We do so because, for us, aesthetics draws attention to dominant ways of sense-making and how things could be otherwise, yet the manner in which we communicate and do research is constitutive of these existing orders. We see value then in exploring different modes of doing research and the alternative modes of sense-making it might make possible (Hawkins, 2020). This is a call, then, for scholarly engagement in disruptive approaches to aesthetics that make other worlds possible (Last, 2017; Nassar, 2021). The emergent field of geopoetics, which bridges radical and creative written poetic work to intervene in the aesthetics of the Earth, might be one reference point here (de Leeuw and Magrane, 2019; Last, 2017, Magrane, 2015), but such a concern can and indeed should span to other forms of creative expression.

Third, there is an opportunity to further align aesthetic thought with fields that pursue emancipatory or transformative goals for cities—redefining what they should be, include, or permit to offer a new "common-sense" on the city to right various injustices. These fields include feminism, post-colonialism, environmentalism, racial justice, queer theory, disability rights, environmental justice, labour movements, climate justice, and anti-capitalism. Of course, such literature certainly already exists in pockets, but aesthetic thought is increasingly inviting such connections. Consider, for instance, Brigstocke and Gassner's (2021: 360) call for a more speculative consideration of aesthetics in the geohumanities "that open up new channels of thought, experience, and practice". At the same time, we see value in investigating "the city" as the stakes of many of these emancipatory fields and this can be deepened and extended.

Take post-colonial thinking on aesthetics, for example, where Édouard Glissant's *Poetics of Relation* (Glissant, 1997) is increasingly popular as a means to offer a disruptive aesthetics (e.g. Brigstocke, 2021; Nassar, 2021; Hoover, 2021). For instance, Nassar draws upon Glissant to ask of Cairo, "are there other ways to story-tell this city without scripting it in an already determined narrative form that fit the character of a Third World city, an Orientalized city, or a city of the Arab Spring" (Brigstocke, 2021: 456). Such post-colonial theory could therefore be applied to recover Indigenous narratives and spaces erased in and around cities, fostering a more inclusive and culturally diverse approach to interpreting what the aesthetics "of" a particular city might be that reflects its historic and contemporary inhabitants' heritage. With this said, there remains a need to think especially "carefully about how we engage the work of thinkers from so-called minority subject-positions" in a manner that respects their work when making such new connections (Jeyasingh, 2024: 6). As Jeyasingh (2024) suggests, perhaps the best option is to read their work in the spirit of opacity rather than transparency. In working against the ideas that the authors' intentions can ever be fully clear or knowable, such a gesture resists the risks of assimilation by instead focusing on the new meanings and connections that emerge through reading.

Notably, there are also other post-colonial thinkers who are less often turned to around aesthetics and the city. Consider, for instance, Achille Mbembe's work on the aesthetics of vulgarity and superfluity which can highlight the display of power through excess and the wastefulness of resources, time, and bodies in cities (Mbembe, 2001; Mbembe, 2004) and Gayatri Chakravorty Spivak's (1985) insights on subalternity and aesthetic education, whose focus on those that exist outside of hegemonic power structures might help scholars develop an ethical engagement for and with the subaltern around cities (Brigstocke and Gassner, 2021). Moreover, while this collection has shown that aesthetics is about more than beauty, imaginaries of the city often still carry assumptions about what is considered beautiful or valuable—frequently rooted in Western, colonial ideals. Post-colonial thought could help critique these frameworks, encouraging a plural understanding of urban aesthetics that challenges the notion of European or Western superiority and embraces diverse forms of beauty that better reflects the diversity that goes into the making of cities. At the same time, a disruptive post-colonial aesthetics around the city might also be specifically positioned against it, with the very notion of the city having been argued to project and enforce colonial relations, serving as "a performative cultural technology of power for monopolizing control over an extended territory" (Woods, 2023: 564). Similar connections can also be drawn with other emancipatory fields beyond this example of post-colonialism, whose agendas are often deeply intertwined with "the city".

Fourth, a more-than-human aesthetics could feature more explicitly in work on the city (Yusoff, 2010; Dixon, 2009; Dixon, Hawkins, and Straughan, 2012). This approach would highlight the interconnections between living and non-living entities involved in shaping the city and shifting focus from human experience and perception to a broader ecological and multispecies framework. Landau-Donnelley and Fridrik in this volume, for instance, considered the connections between humans, more-than-humans, and aesthetic expressions—such as paint on walls or weeds on structures. Moreover, more-than-human aesthetics could be used to explore how environmental art and installations within cities, which engage with or respond to natural processes, prompt viewers to consider their roles in broader ecosystems. Post-human sensibilities could be incorporated by considering the limitations of human-centric thinking, especially as technological, artificial, or algorithmic systems mediate perception and experience. Such an approach would raise critical questions about the boundaries between humans and machines, a topic touched upon by Degen and Gujarro in their discussion of digitally mediated urban branding.

Fifth, and as somewhat of a unifying thread for the points made thus far, attention should be directed towards the relationship between aesthetics, the city, and negativity (Bissell, Rose, and Harrison, 2021). We do not refer to negativity in the sense of pessimism, but a concern for "that which is not"—for absences, exclusions, and non-existences (Bissell, Rose, and Harrison, 2021). If one accepts that a key purpose of urban studies is to improve the urban condition, and that the ways in which the city is made sense of are central to such a concern, then a key question arises for aesthetics: what are the negative aesthetics of cities? This question encourages one to confront the blind spots, exclusions, and alternatives inherent in common-sense understandings of cities and sensory experiences of them. It calls for an exploration of how these relate to power geometries and lived experiences.

Thinking about the negative aesthetics of cities, then, invites an interrogation of how the ways cities are made sense of—and the various violences, forms, relations, atmospheres, and affects that relate to them—might become otherwise. Scholars might also consider how alternative ways of making sense of the city are resisted—be that through forms of strategic ignorance that serve certain interests (Blakey, 2021), or through anaesthetic tactics to resist certain affective relations that are set in motion around the city (Bissell, 2022). With this in mind, the task of a negatively attuned approach to aesthetics and the city might be around both actively attending to alternatives and/or creating the conditions for such alternatives to emerge. In other words, it is about ensuring a strong emancipatory inflexion to our work and an appreciation for the contingency of existing modes of sense-making (Landau-Donnelly and Pohl, 2023). It is owing to this commitment that our ways of making sense of the city can always be different that drives our recommendation for an agonistic approach to aesthetics, and by extension, the city, avoiding definitive theorisations and positioning them as subjects of ongoing reflexive—yet productive—disagreement and inquiry.

Finally, as these suggestions are taken forward, we want to end this volume with a warning: the words shared in the pages of this book should not be taken as a hollow defence of the city or an invitation to engage in unreflexive discussions about it. The concepts that scholars use—whether in designing or doing fieldwork or when interpreting findings—can shape what is knowable and thinkable. Occasionally, these concepts can also become overly prescriptive, interrupting our ability to see the world as otherwise. As the introduction outlines, the concept of the city in particular has raised significant scholarly concern, with the emergence of critiques around city-centric thought and methodological cityism. Nevertheless, when approached with caution and reflexivity, aesthetics can illuminate how the city shapes urban life—or even how it may not. The key epistemological point raised in this book is that aesthetics, rather than the city, should serve as the starting point for analysis. This shift allows an exploration of the diverse ways in which the city may or may not be present in everyday life, discourse, policy, and beyond.

There is no one-size-fits-all approach to aesthetics. It is the diversity of approaches, neighbouring concepts, and underpinning theoretical persuasions that enables a rich and critically reflective dialogue about the city and how it is made sense of within urban studies. We have not sought to provide a definitive definition of aesthetics or the city; both are open and contestable concepts influenced by the theories and positionalities from which they are known. Rather, we have presented a fertile space for critical thought: an opportunity for scholars from various disciplines and theoretical interests to explore the role of the city in urban life. In this regard, we have advocated an agonistic approach to deploying aesthetics in relation to the city that resists totalising narratives and values disagreement. Doing so, we argue, allows us to understand and perhaps even subvert or unseat the role of "the city" in relation to urban experience. As the contributions in this volume have demonstrated, it is the diverse ensemble of different approaches to aesthetics that allows us to make sense of this thing we call "the city".

References

Anderson, B., (2019). Cultural geography II: The force of representations. *Progress in Human Geography*, 43(6), pp. 1120–1132.
Angelo, H. and Wachsmuth, D., (2015). Urbanizing urban political ecology: A critique of methodological cityism. *International Journal of Urban and Regional Research*, 39(1), pp. 16–27.
Arnold, E., (2019). Aesthetic practices of psychogeography and photography. *Geography Compass*, 13(2), p.e12419.
Barron, A., (2021a). More-than-representational approaches to the life-course. *Social & Cultural Geography*, 22(5), pp. 603–626.
Barron, A., (2021b). The taking place of older age. *Cultural Geographies*, 28(4), pp. 661–674.
Barry, A., Born, G. and Weszkalnys, G., 2008. Logics of interdisciplinarity. *Economy and Society*, 37(1), pp. 20–49.
Blakey, J. (2021). Accounting for elephants: The (post) politics of carbon omissions. *Geoforum*, 121, pp. 1–11.
Blakey, J., Machen, R., Ruez, D. and García, P.M., (2022). Intervention: Engaging post-foundational political theory requires an 'enmeshed'approach. *Political Geography*, 99, p. 102689.
Bissell, D., Rose, M. and Harrison, P. (Eds.) (2021). *Negative Geographies: Exploring the Politics of Limits.* University of Nebraska Press.
Bissell, D. (2022). The anaesthetic politics of being unaffected: Embodying insecure digital platform labour. *Antipode*, 54(1), pp. 86–105.
Brenner, N., (2019). *New Urban Spaces: Urban Theory and the Scale Question.* Oxford University Press.
Brigstocke, J., (2021). The aesthetics of sand: Reclaiming Hong Kong's unsettled grounds. *GeoHumanities*, 7(2), pp. 370–390.
Brigstocke, J. and Gassner, G., (2021). Materiality, race, and speculative aesthetics. *GeoHumanities*, 7(2), pp. 359–369.
de Leeuw, S. and Magrane, E., 2019. Geopoetics. *Keywords in Radical Geography: Antipode*, 50, pp. 146–150.
Dixon, D.P., (2009). Creating the semi-living: on politics, aesthetics and the more-than-human. *Transactions of the Institute of British Geographers*, 34(4), pp. 411–425.
Dixon, D., Hawkins, H. and Straughan, E., (2012). Of human birds and living rocks: Remaking aesthetics for post-human worlds. *Dialogues in Human Geography*, 2(3), pp. 249–270.
Gandy, M., (2021). Film as method in the geohumanities. *GeoHumanities*, 7(2), pp. 605–624.
Ghertner, D.A., (2015). *Rule by aesthetics: World-class city making in Delhi.* Oxford: Oxford University Press.

Glissant, É., (1997). Poetics of Relation. Michigan: University of Michigan Press.
Hawkins, H., (2015). Creative geographic methods: knowing, representing, intervening. On composing place and page. *Cultural Geographies*, 22(2), pp. 247–268.
Hawkins, H., (2021). Cultural geography I: mediums. *Progress in Human Geography*, 45(6), pp. 1709–1720.
Hawkins, H., (2020). *Geography, Art, Research: Artistic Research in the GeoHumanities*. Routledge.
Hawkins, H. and Straughan, E., (2016). Introduction: For geographical aesthetics. In *Geographical Aesthetics*, edited by H. Hawkins and E. Straughan. London: Routledge. pp. 1–17.
Hitchings, R., (2012). People can talk about their practices. *Area*, 44(1), pp. 61–67.
Hoover, E.M., (2021). Poetic commoning in European cities–Or on the alchemy of concrete. *GeoHumanities*, 7(2), pp. 464–474.
Jeyasingh, T.E. (2024). Édouard Glissant and the importance of reading well: Opacitic-reading as geographic method. *Transactions of the Institute of British Geographers*, pp. 1–15. https://doi.org/10.1111/tran.12685
Landau-Donnelly F. and Pohl, L., (2023). Towards a post-foundational geography: Spaces of negativity, contingency, and antagonism. *Progress in Human Geography*, 47(4), pp. 481–499.
Last, A. (2017). We are the world? Anthropocene cultural production between geopoetics and geopolitics. *Theory, Culture & Society*, *34*(2–3), pp. 147–168.
Magrane, E., 2015. Situating geopoetics. *GeoHumanities*, 1(1), pp. 86–102.
Marchart, O. (2019). *Conflictual Aesthetics: Artistic Activism and the Public Sphere*. Berlin: Sternberg Press
Mason, O. and Riding, J., (2023). Reimagining landscape: Materiality, decoloniality, and creativity. *Progress in Human Geography*, *47*(6), pp. 769–789.
Mbembe, A., (2001). *On the Postcolony* (Vol. 41). Univ of California Press.
Mbembe, A., (2004). Aesthetics of superfluity. *Public Culture*, *16*, pp. 373–406.
Mouffe, C. (2005). *On the Political*. London: Routledge.
Murakami Wood, D. and Abe, K., (2011). The aesthetics of control: Mega events and transformations in Japanese urban order. *Urban Studies*, 48(15), pp. 3241–3257.
Nassar, A., (2021). Geopoetics as disruptive aesthetics: Vignettes from Cairo. *GeoHumanities*, 7(2), pp. 455–463.
Rancière, J., (1999). *Disagreement: Politics and Philosophy*. Minnesota: University of Minnesota Press.
Spivak, Gayatri Chakravorty. (1985). "Can the subaltern speak? Speculations on widow sacrifice." *Wedge* 7, no. 8 (Winter/Spring): 118–131.
Von Benzon, N., Holton, M., Wilkinson, C. and Wilkinson, S. (2023). *Creative Methods for Human Geographers*. London: SAGE Publications.
Woods, M., (2023). Against the City: What Derek Walcott has to teach us about the city imaginary. *GeoHumanities*, 9(2), pp. 554–568.
Yusoff, K., (2010). Biopolitical economies and the political aesthetics of climate change. *Theory, Culture & Society*, 27(2–3), pp. 73–99.

Index

FSC
www.fsc.org
MIX
Papier | Fördert gute Waldnutzung
FSC® C083411

Kolibri 360 GmbH
Ferdinand-Jühlke-Straße 7
99095 Erfurt, Deutschland
produktsicherheit@kolibri360.de